Advances in Semiconductor Physics, Devices and Quantum Dots - Nanotechnology and Future Challenges

Edited by Tingting Yin, Jean-Luc Autran and Daniela Munteanu

Published in London, United Kingdom

Advances in Semiconductor Physics, Devices and Quantum Dots - Nanotechnology and Future Challenges
http://dx.doi.org/10.5772/intechopen.111359
Edited by Tingting Yin, Jean-Luc Autran and Daniela Munteanu

Contributors
Chengbin Kang, César Camas, Daniela Munteanu, Ehab A. Okba, El-Zeiny M. Ebeid, Hansong Zhang, Heber Vilchis, Jean-Luc Autran, Jorge Conde, Keran Jiao, Lawrence B. Crowell, Lemi Demeyu, Mingjie Li, Ning Tu, Pengfei Song, S.W. Ricky Lee, Seyfan Kelil Shukri, Xiaowu Gao, Yongjie Wang, Ziwei Zhao

First published in London, United Kingdom, 2025 by IntechOpen
IntechOpen is the global imprint of INTECHOPEN LIMITED, registered in England and Wales, registration number: 11086078, 167-169 Great Portland Street, London, W1W 5PF, United Kingdom

For EU product safety concerns: IN TECH d.o.o., Prolaz Marije Krucifikse Kozulić 3, 51000 Rijeka, Croatia, info@intechopen.com or visit our website at intechopen.com.

British Library Cataloguing-in-Publication Data
A catalogue record for this book is available from the British Library

Advances in Semiconductor Physics, Devices and Quantum Dots - Nanotechnology and Future Challenges
Edited by Tingting Yin, Jean-Luc Autran and Daniela Munteanu
p. cm.
Print ISBN 978-0-85466-512-9
Online ISBN 978-0-85466-511-2
eBook (PDF) ISBN 978-0-85466-513-6

Meet the editors

Dr. Tingting Yin has dedicated almost 10 years to optical spectroscopy of low-dimensional semiconductors under different external stimuli (high pressure, low temperature, and strong magnetic field), particularly focusing on spectral demonstration, mechanism exploration, and the multi-field manipulation of exciton properties. Dr. Yin has an interdisciplinary research background encompassing optical spectroscopy, high-pressure physics, and materials science. Dr. Yin has achieved a significant impact in various research fields, and the outcomes have been documented in high-impact international journals. Dr. Yin was awarded the Presidential Postdoctoral Fellowship at Nanyang Technological University. In 2023, Dr. Yin was named a Global Young Scientist for 2024.

Prof. Jean-Luc Autran is Distinguished Professor of Physics at the University of Rennes and Honorary Member of the University Institute of France (IUF). He currently serves as the Head of the Institute of Physics of Rennes (IPR, UMR 6251). From 2000 to 2023, he served as Distinguished Professor at Aix-Marseille University (AMU). He also held the position of Head of the Institute for Materials, Microelectronics and Nanosciences of Provence (IM2NP, UMR 7334) from 2018 to 2023 and Director of the "Radiation Effects and Electrical Reliability" (REER) Joint Laboratory between IM2NP and STMicroelectronics from 2014 to 2020. With 30 years of experience in the field of semiconductor interface defects, physics of advanced CMOS devices, and radiation effects in microelectronics, his current research interests are focused on the physics of single-event effects. This includes characterization, analytical modeling, and numerical simulation. Jean-Luc Autran has authored or co-authored more than 350 papers that have been published in international journals and conferences. He co-authored the book "Soft Errors: From Particles to Circuits" with Daniela Munteanu.

Dr. Daniela Munteanu is Director of Research at the National Center for Scientific Research (CNRS). She was a member of the Institute for Materials Microelectronics and Nanoscience of Provence (IM2NP, UMR 7334), and she recently joined the Institute of Physics in Rennes (IPR, UMR 6251). Daniela Munteanu has 25 years of experience in the domains of characterization, modeling, and simulation of semiconductor devices. Her research focuses on emerging CMOS devices, compact modeling, numerical simulation, and radiation effects in nanoelectronics, with a particular interest in single events (soft errors) induced by particles from both natural and artificial radiation environments. Daniela Munteanu is the author or co-author of more than 250 papers published in international journals and conferences. She has supervised 15 Ph.D. theses and co-authored with Jean-Luc Autran the book "Soft Errors: From Particles to Circuits".

Contents

Preface

This book is a collection of chapters about developments in the field of semiconductor physics, devices, and quantum dots.

The book includes scholarly contributions by various authors and was edited by a group of experts in the field. Each chapter is complete in itself; however, each chapter is also directly related to the book's topics and objectives.

The book is divided into two sections: "Quantum Matter" and "Advanced Semiconductors".

The target audience comprises scholars and specialists in the relevant field.

Jean-Luc Autran
Institute of Physics of Rennes,
University of Rennes,
Rennes, France

Daniela Munteanu
Institute of Physics of Rennes,
National Center for Scientific Research (CNRS),
Rennes, France

Tingting Yin
Nanyang Technological University,
Singapore

Section 1

Quantum Matter

Chapter 1

Quantum Dots: Their Unique Properties and Contemporary Applications

El-Zeiny M. Ebeid and Ehab A. Okba

Abstract

This chapter describes the exciton concept and exciton Bohr radius. It describes new and diverse QDs applications in the fields of photonics, quantum dot lasers, photon upconversion (PUC) and downconversion (PDC) and their applications, biosensors, environmental sensing, ratiometric fluorescence and colorimetric dual-mode sensors, food quality sensing, cancer biomarkers detection, non-photonic medical imaging including magnetic resonance imaging (MRI), radiolabeled quantum dots, positron emission tomography (PET), drug delivery, blood-brain barrier (BBB) crossing, electrochemical sensing, photocatalysis including CO_2 reduction, H_2 production, and environmental remediation. The chapter ends with a Conclusion and prospects section expecting crucial QDs industrial applications such as displays, solar cells, wastewater treatment, quantum computers, and biomedical applications. Heavy metal-free QDs formulations are a demand to minimize traditional QDs toxicity. There is progress in using non-toxic and eco-friendly starting materials, including carbon-based, biomolecules-based, silicon-based, and ternary I-III-VI QDs alternatives.

Keywords: QD lasers, photon upconversion (PUC), photon downconversion (PDC), medical imaging, drug delivery, blood-brain barrier, biosensors, environmental sensing, electrochemical sensing, food quality, photocatalysis

1. Introduction

Quantum dots (QDs) are nanoscale particles that exhibit unique optical and electronic properties due to quantum confinement effects. They are characterized by small sizes that are smaller than twice the size of its exciton Bohr radius. The principles behind quantum dots are rooted in quantum mechanics, solid-state physics, and nanotechnology. The confinement of electrons and holes within the quantum dot structure leads to the quantization of energy levels. In other words, the energy levels of electrons and holes become discrete rather than continuous.

Quantum confinement in QDs leads to optical tunability based on their size. Some unique spectral properties of QDs are relatively narrow and symmetrical emission peaks spanning the UV to the IR spectral regions, high emission quantum yields, high photostability, high molar absorptivities, long emission lifetimes, and large Stokes shifts. This led to a variety of important research and commercial applications including bioimaging, solar cells, LEDs, diode lasers, transistors, phosphor displays, electrochemiluminescence, and on/off switches on electrode surfaces [1–5].

The small QDs size makes them suitable for biological and medical applications including bioimaging and biosensors. For instance, QDs are used in studying intracellular processes, targeting tumors, real-time observation of cell trafficking in vivo, diagnosis, and cellular imaging at high resolutions [6]. The large volume-to-size ratio renders QDs of valuable applications in catalysis, solar cells, delivery of therapeutics, and highly enhanced adsorption of chemicals [7].

More recently, electrochemical nanosensors in the form of carbon quantum dots (CQDs), graphene quantum dots (GQDs), and semiconductor quantum dots are widely applied to fabricate sensing systems [8]. This assured better redox properties allowing detection of many analytes such as ions, pharmaceuticals, small molecules, and biological macromolecules.

The high edge-to-area ratio in some QDs leads to spin polarized edge states, which can generate magnetic properties. Magnetic quantum dots (MQDs) find applications in catalysis, sensors, cancer treatment, environmental remediation, and contrast-improving agents in magnetic resonance imaging (MRI) [9].

With the number of drug-resistant infections continually rising, QDs find applications to treat antibiotic-resistant infections by producing reactive oxygen species (ROS) when exposed to light, which can result in oxidative damage to bacterial cells [10], or alternatively producing antibiotics that release a superoxide enzyme with quantum dots [11, 12].

2. Photonic applications of QDs

2.1 Quantum dot lasers (QD lasers)

QD lasers apply semiconductor QDs as active laser media in their light-emitting regions. QD lasers avoid some of the drawbacks associated with traditional semiconductor lasers based on bulk or quantum well (Q-well) lasers [13, 14]. In Q-well lasers, a change in the dimensions of the well changes the output wavelength, and a larger population inversion is needed for laser action to be achieved. In QD lasers, a wide range of light emission is obtained by controlling dot size, and small active volumes are needed, allowing lasing action using smaller population inversion. This leads to less temperature dependence of threshold current density, an increase in gain (2–3 times larger than Q-well), high-frequency operation, and more laser efficiency [15].

Due to the discrete nature of energy levels in QDs, QD semiconductor lasers are characterized by properties such as excellent temperature stability, tolerance to crystalline defects, low threshold current density, and reduced reflection sensitivity. They can produce wavelengths from the near-infrared (NIR) region to the visible and even into the ultraviolet (UV) range. Typical applications often operate within the near-infrared spectrum, employing wavelengths ranging from 600 to 1600 nm. The discrete nature also serves in quantum cryptographic communication as well as quantum information processing for applications such as quantum computers. The use of discrete energy level of quantum dots allows the absorption of long-wavelength light necessary for future development of high-efficiency solar cells. A number of quantum dots applications include high-sensitivity, low-dark-current infrared detectors

DOI: http://dx.doi.org/10.5772/intechopen.1005582

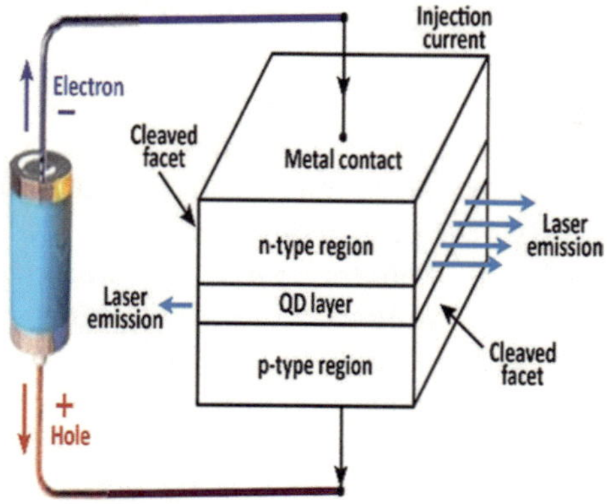

Figure 1.
The construction of a quantum dot laser diode with p- and n-region layers. The active layer situated between them contains quantum dots. Metal contacts connected to the p-region and n-region facilitate electrical current flow within the diode.

allowing a strong carrier confinement [16]. The construction of a quantum dot laser diode is shown in **Figure 1**.

Wu et al. developed a theory of output optical power of QD lasers based on carrier injection from the cladding layers to the optical confinement layer (OCL) followed by carrier capture from the OCL into the QD excited state and carrier escape from the QD excited state to the OCL. Spontaneous radiative recombination in the OCL occurs with downward transition in QDs (intradot relaxation), upward transition in QDs, spontaneous and stimulated radiative recombination via QDs excited state, and spontaneous and stimulated radiative recombinations via the ground state in QDs [17]. The faces of p- and n-layers are coated with high reflectivity materials causing light reflection back and forth to increase stimulated emission and in turn, the laser emission is enhanced (**Figure 2**).

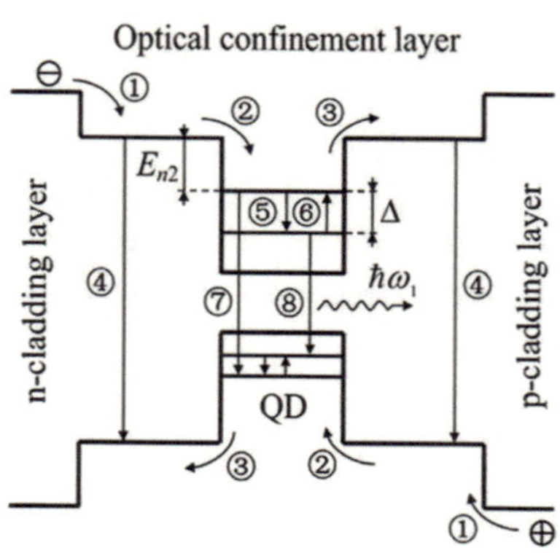

Figure 2.
Energy band diagram of a QD laser (the layers are not drawn to scale). The main processes (shown by arrows) are as follows: ① carrier injection from the cladding layers to the optical confinement layer (OCL), ② carrier capture from the OCL into the QD excited state, ③ carrier escape from the QD excited state to the OCL, ④ spontaneous radiative recombination in the OCL, ⑤ downward transition in QDs (intradot relaxation), ⑥ upward transition in QDs, ⑦ spontaneous and stimulated radiative recombination via the excited state in QDs, and ⑧ spontaneous and stimulated radiative recombinations via the ground state in QDs. Adapted with permission from Ref. [17]. Copyright (2015), AIP Publishing.

2.2 QDs in photon upconversion (PUC) and downconversion (PDC) and their applications

Photon upconversion (PUC) refers to a nonlinear optical process whereby two or more low-energy photons are converted into one high-energy photon (i.e., anti-Stokes behavior) [18]. The discovery of photon upconversion was coined in 1966 by François Auzel who showed that an infrared photon of light could be upconverted into a photon in the visible-light region in ytterbium-erbium and ytterbium-thulium systems [19].

On the other hand, photon downconversion (PDC) is a process in which a high-energy ultraviolet (UV) photon is converted into a low-energy visible or NIR photon. The (PDC) phenomenon was first reported in YF_3: Pr^{3+}, in which one ultraviolet (UV) photon was converted into two visible (Vis.) photons of lower energy [20]. Lanthanide ions dopants are applied as activator ions since they have multiple $_4f$ excitation levels and filled $_5s$ and $_5p$ shells. These outer-filled shells act as a shield for the $_4f$ electrons, thus producing sharp f-f transition bands. The electronic configuration of Lanthanides is [Xe] $_4f^{1-14}$ $_5d^{0-10}$ $_6s^2$ where the $_4f$ electrons are not affected by environmental conditions due to outer orbits $_5d^{0-10}$ $_6s^2$ electrons. The coupling of the $_4f$ electronic excited states to the surrounding lattice is weak and this leads to long excited state lifetimes and sharp optical line shapes [21].

Both (PUC) and (PDC) phenomena are helpful in many fields, including photonics, nanomedicine, biomedical, and therapeutic applications. **Table 1** summarizes some common methods to get (PUC) and **Table 2** summarizes some common methods to get (PDC) phenomena [22].

A brief description of processes in **Tables 1** and **2** is given as follows:

2.2.1 Photon avalanche (PA)

In photon avalanche (PA), multiple photons combine to produce a single higher energy photon. This phenomenon was first discovered by Chivian in 1979 in Pr^{3+} doped $LaCl_3/LaBr_3$ systems based on infrared quantum counters [22–26]. Although PA is seen in some systems, it is the least observed mechanism for upconversion [26].

2.2.2 Cooperative sensitization upconversion (CSU)

Cooperative sensitization upconversion implies the simultaneous decay of two lanthanide ions in their excited states to their ground states, generating a higher energy photon [22, 27, 28]. This encounters one or more elementary steps (sensitization or luminescence). Similarly, in cooperative luminescence, two excited state ions transfer their energy to a neighboring ion in one elementary step.

2.2.3 Excited state absorption (ESA)

In excited-state absorption, a previously exited electron within an ion absorbs a photon [29] and is promoted by long-lived metastable energy levels. This exists in many lanthanides including Er^{3+}, Tm^{3+}, and Ho^{3+} ions.

2.2.4 Energy-transfer upconversion (ETU)

In energy-transfer upconversion, energy transfers from one exited ion to another. An ideal energy-mediated upconversion requires a core-shell nano-system with a

DOI: http://dx.doi.org/10.5772/intechopen.1005582

Photon upconversion (PU)							
						Two-photon absorption (TPA)	
Photon avalanche (PA) (**Figure 3a**)	Cooperative sensitization upconversion (CSU) (**Figure 3b**)	excited state absorption (ESA) (**Figure 3c**)	Energy-transfer upconversion (ETU) (**Figure 3d**)	Triplet-triplet annihilation (TTA) (**Figure 3e**)	Energy migration-mediated upconversion (EMU) (**Figure 3f**)	Simultaneous Virtual intermediate states (dotted) (**Figure 3g**)	Sequential Real Metastable excited states. UC emission is much more efficient (**Figure 3h**)

Table 1.
Some common methods for (PUC) phenomenon.

Photon downconversion (PDC)		
Quantum cutting (QC) phenomenon where one high-energy photon is converted into two or more than two photons. 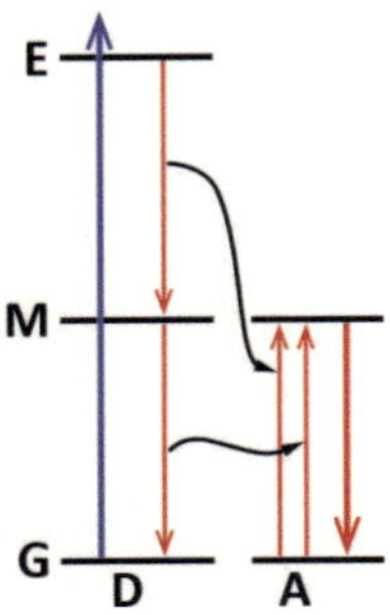	Near-infrared quantum cutting long persistent luminescence (NQPL) 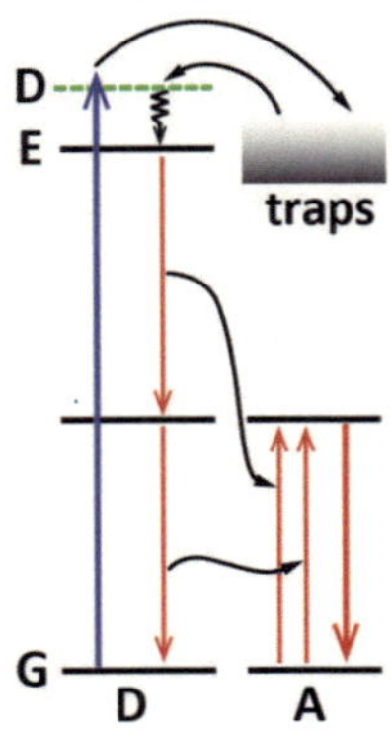	Long persistent luminescence (LPL). 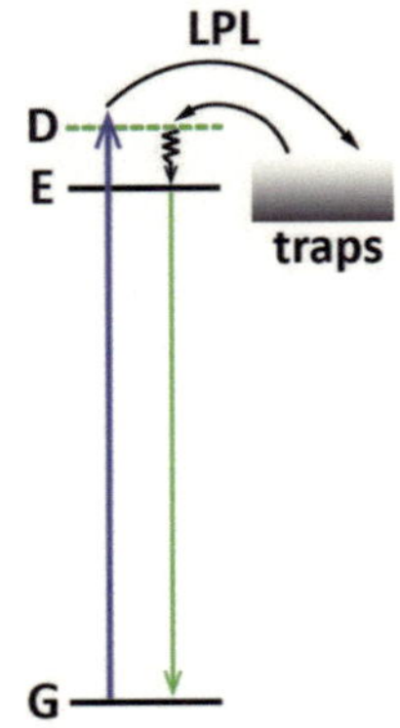

Table 2.
Some common methods for (PDC) phenomenon.

sensitizer ion (e.g., Yb^{3+}) embedded in the core. Its role is to transfer energy via a ladder (e.g., Tm^{3+}) to a migratory ion capable of promoting the energy to an activator ion located in the shell. Energy-transfer upconversion (ETU) and excited state absorption (ESA) are the two most common processes by which upconversion occurs in lanthanide-doped nanoscale materials [20, 30].

2.2.5 Simultaneous two-photon absorption (TPA)

Simultaneous two-photon absorption involves virtual intermediate states for absorption of two or more photons [22].

2.2.6 Sequential two-photon absorption (TPA)

In sequential two-photon absorption, real metastable excited states allow for sequential absorption. Therefore, a necessary condition for upconverting systems is the existence of optically active long-lived excited states. It is of more interest compared to simultaneous two-photon absorption that is based on virtual energy levels. In sequential two-photon absorption, lower energy densities are required (low-power and incoherent excitation sources) [20, 22].

2.2.7 Triplet-triplet annihilation (TTA)

Triplet-triplet annihilation (TTA)-based UC emission is common in organic molecules. Since TTA systems are usually very sensitive toward oxygen (a known triplet quencher), special experimental precautions are required. Inorganic semiconductor quantum dots (QDs) are potential candidates in TTA showing a potential role in harvesting low-energy photons, which can then be used for photon upconversion (PUC), via triplet-triplet annihilation (TTA) [31].

Wang et al. achieved a strong coupling regime in which excited carriers spatially delocalize across anthracene molecules strongly linked to silicon quantum dots via

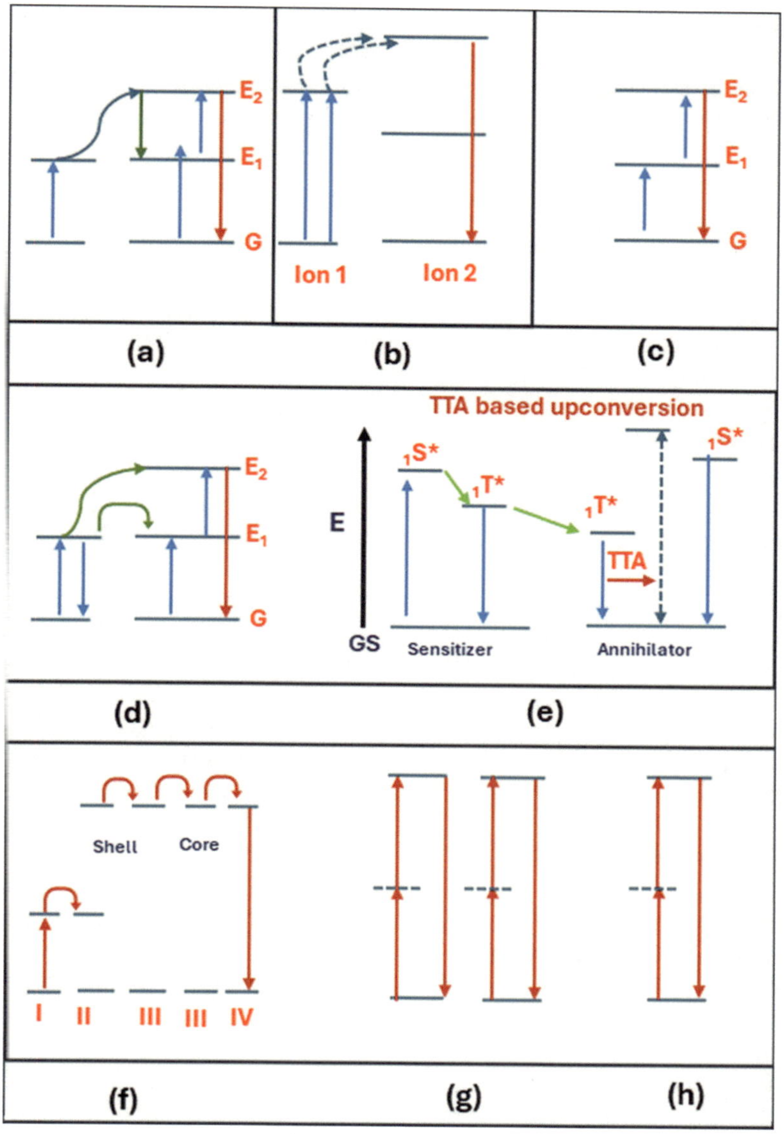

Figure 3.
Some common methods for (PUC) phenomenon given in ***Table 1****.*

double bonds. Triplet exciton annihilation leads to photon upconversion with high efficiency (17.2%) and low threshold intensity (0.5 W cm^{-2}) [32].

2.2.8 Energy migration upconversion (EMU)

Energy migration upconversion (EMU) is a complex process mainly observed in core-shell structures in which four kinds of lanthanide ions are embedded. It was first proposed by Liu and coworkers in 2011 [33]. A sensitizer ion (type I) located in the shell is first excited and then transfers its excitation energy to an accumulator ion (type II) that is also located in the shell. The energy transfers from the high-lying excited state of the accumulator to a migrator ion (type III) that is also located in the shell. An excited migrator ion then forwards its excitation energy across the core-shell interface via random hopping through many (type III) migrator ions that populate the core-shell interface and the core. Finally, the migrating energy is trapped by the activator ion (type IV) located in the core, resulting in upconverted emission [34–36].

2.2.9 Photon downconversion (PDC)

2.2.10 Quantum cutting (QC) phenomenon

In quantum cutting (QC) phenomenon, a high-energy photon is converted into two or more than two photons of lower energy. It was long applied to get photon downconversion (PDC) [37, 38].

2.2.11 Long persistent luminescence (LPL)

Long persistent luminescence (LPL) is a phenomenon in which light emission lasts for a very long time (hours) after excitation is ceased. Referring to **Table 2**, an ion is excited to a delocalized excited state (D). The delocalized electrons can be trapped for a long time. After ceasing the excitation sources, the trapped electrons can be de-trapped back to the delocalized state (D) because of thermal or photon activation. This is followed by a non-radiative relaxation pathway to the E state and then radiatively deactivates to the ground state (G) giving LPL. Representative LPL systems include $SrAl_2O_4{:}Eu^{2+}$, Dy^{3+} (green), $CaAl_2O_4{:}Eu^{2+}$, Nd^{3+} (blue), $Y_2O_2S{:}Eu^{3+}$, Mg^{2+}, Ti^{4+} (red), and $Zn_3Ga_2Ge_2O_{10}{:}Cr^{3+}$ (near-infrared: NIR). Near-infrared quantum cutting long persistent luminescence (NQPL) is also a known phenomenon [38].

2.2.12 QDs in photonic technology

In the field of photonic technology, quantum dots (QDs) have become an essential component, greatly impacting the advancement of high-performance devices, especially in the LED and QLED markets. Semiconductor quantum dots (QDs) have many useful features that can be used in photonic applications. These include the ability to modify the bandgap, structural stability, cost-effectiveness, vast area coverage, and solution processability. Quantum dots (QDs) are essential for producing vivid, dynamic displays with high color accuracy, and are used in QLEDs for efficient light production. Their adaptable properties have a significant impact on various disciplines, including soft robotics, computer hardware, and medical devices [39].

Perovskite quantum dots (QDs) have greatly enhanced the performance and efficiency of light-emitting diodes (LEDs), especially in the case of QLEDs. The nanocrystals, namely the all-inorganic cesium lead halide ($CsPbX_3$) QDs, exhibit exceptional photoelectric characteristics, rendering them very suitable for photonic applications such as LEDs. The utilization of titanium-doped cesium lead iodide quantum dots (QDs) in conjunction with yellow-emitting phosphors, blue LED chips demonstrates the promise of perovskite QDs in solid-state lighting and display technologies [40–47].

2.2.13 Photon upconversion (PU) and photon downconversion (PDC) applications

He et al. reported a novel fluorescent probe based on the combination of rare earth upconversion QDs, perovskite quantum dots, Molecular Beacon as a carrier, and Black Hole Quencher-1. The probe was used for cancer early detection via specific recognition of miRNA-155 at concentrations down to a detection limit of 73.5 pM [48, 49].

Lai and Wu reported that interfacial triplet energy transfer from red-emitting CdSe QDs to surface-anchored Rhodamine B is enhanced upon coating with ZnS shell. The generated RB triplet state can be applied in singlet oxygen generation for the purpose of application in photodynamic therapy (PDT) [50].

Lu et al. reviewed the potential applications of QDs in optical energy conversion into other useful forms of energy including optical energy to electricity, photon upconversion (PUC) and photon downconversion (PDC), optical energy to chemical bonds including green hydrogen generation, QDs with cocatalysts, integrating QDs into photoelectrochemical cells, as photocatalysts in organic synthesis, CO_2 reduction and N_2 fixation [50].

The demand for non-toxic QDs is growing. Most existing quantum dots contain toxic metals such as cadmium and lead, which limit their electronic and medical applications. Safer core-shell quantum dots of InP core surrounded by a thin shell of ZnS that emit pure blue light in a solid state and solution were prepared. The dots emit blue light even inside the biological cells [51].

2.3 Quantum dots as sensors

Quantum dots (QDs) have emerged as promising tools in the field of medical imaging, offering unique advantages that can revolutionize diagnostics and treatment monitoring. Their size-tunable fluorescence emission, high photostability, and brightness make them ideal candidates for various imaging modalities. In fluorescence-based imaging, QDs outperform conventional organic dyes, providing enhanced image quality and reduced photobleaching. Because of their surface functionalization capabilities and exceptional optical qualities, quantum dots (QDs) have in vivo applications that allow for the real-time imaging of biological processes [52, 53].

Quantum dots probes enjoy unique properties that make them excellent fluorescent imaging agents. These include:

1. Their size-modulated absorbance and emission enable the fabrication of identically constructed probes with diverse optical characteristics for the aim of multiplexing [54, 55].

2. High photostability allows imaging for lengthy periods of time without signal loss—potentially useful for fluorescence-guided surgery [55].

3. A large Stokes shift makes it easier to block out the excitation light and allows the use of a single excitation wavelength for imaging in multiple colors [54].

4. Their high brightness is attributed to their high excitation cross-section values. This allows improved sensitivity and allows fluorescence imaging at low receptor concentration levels [55].

5. Despite having similar photoluminescence quantum yields (PLQYs) in the UV-visible region, QDs have superior PLQYs application in the NIR region compared to organic dyes [56].

6. The remarkably long lifetimes of excited states allow the development of probes with time-delay microscopy and this avoids interference with the autofluorescence of cells and other moieties [57].

7. High-surface area-to-volume ratio in QDs allows for efficient functionalization with other imaging agents and the creation of multi-modal probes to be used in imaging of disease states at different scale lengths and different tissue depths [58].

The unique optical characteristics and multifunctionality of nanoscale semiconductor particles, such as carbon-based, chalcogenide, black phosphorus, and cadmium containing QDs, make them valuable instruments for imaging cancer cells. Benefits of QDs include intense fluorescence, a wide excitation range, narrow emission, and excellent resistance to photobleaching. QDs are intriguing prospects for enhancing cancer cell imaging in medical research and diagnostics because of their potential for early disease identification and their ability to be adjusted for targeted imaging [59].

2.3.1 Biosensors

CQDs are utilized as biosensor carriers due to their excellent water solubility, surface modification flexibility, non-toxicity at the applied concentrations, excitation-dependent emission in different light regions, high biocompatibility, good cell permeability, and strong photostability. These CQD-based biosensors have been employed for real-time monitoring of various analytes, including glucose [60], cellular copper ions [61], phosphate ions [62], iron ions [63], potassium ions [64], pH [65], and nucleic acids [66]. In nucleic acid detection, CQDs are used as efficient fluorescent sensing platforms capable of detecting single-base mismatches. This approach involves the initial adsorption of fluorescently labeled single-stranded DNA probes onto CQDs through π-π interactions, leading to significant fluorescence quenching. Upon specific hybridization with their target sequences to form double-stranded DNA, the dsDNA detaches from the CQD surface, resulting in fluorescence recovery, enabling the detection of the target DNA [62].

Quantum dots (QDs) have emerged as one of the most cost-effective and promising indicators for viral infections. Leveraging QDs' ultrasensitive capabilities as biological probes, combined with the direct use of physiological components, allows for the simultaneous detection of several viral indicators. The effectiveness of QDs as detection techniques for human pandemic viruses is based on incorporating quantum dot-based biomarkers into ultrasensitive probes for multiplex detection of human viral infections. This collaborative approach has great promise for improving our ability to combat viral infections effectively [67].

Quantum dots (QDs) have the unique capacity to selectively attach to biomolecules, including oligonucleotides, proteins (including peptides, antibodies, or enzymes), and polymers, due to their special characteristics. The flexibility of this technology allows for a wide range of sensing applications, as demonstrated by the quick and accurate detection of avian influenza (AIV) virus, namely the H5N1 subtype, using fluoroimmunoassay assays based on quantum dots (QDs) [68].

2.3.2 Cancer biomarkers detection

Fluorescent quantum dots are utilized in the identification of liver cancer biomarkers using aptamer-based fluorescence biosensors, which are crucial for the diagnosis of liver cancer. These biosensors employ aptamers to label fluorescent CdTe quantum dots. Aptamers, being oligonucleotides composed of either DNA or RNA, possess the capacity to selectively bind to certain biomolecules, including alpha-fetoprotein (AFP), a prominent cancer biomarker. Fluorescence resonance energy transfer (FRET) takes place when quantum dots get close to gold nanoparticles that have been modified with an anti-AFP antibody. This action results in the reduction of the fluorescence signal emitted by the quantum dots. Employing this method enables accurate and dependable assessment of AFP levels, which is essential for both the identification and ongoing surveillance of liver cancer [69]. FRET is a distance-dependent interaction between light-sensitive molecules, transferring energy non-radiatively between donor and acceptor molecules. It is efficient when within 1–10 nm, making it useful for molecular interactions and conformational changes, and is used in biological and chemical sensing. Metal-enhanced fluorescence (MEF) is a technique that uses metallic nanoparticles to enhance fluorescence intensity in fluorophores. This interaction increases excitation rate and lifetime, enhancing fluorescence sensitivity and detection limits in fluorescent-based assays, especially for biomolecule detection in diagnostic applications.

2.3.3 Quantum dots in environmental sensing

QDs facilitate the detection of pollutants and toxic substances at very low concentrations, contributing to improved environmental protection and safety. The role of QDs in sensing is pivotal, leveraging their exceptional optical characteristics to enhance the performance of sensors in terms of sensitivity, specificity, and multiplexing capabilities, thus meeting the demands of various complex sensing tasks. Si QDs exhibited good water solubility and high stability. Under the optimized conditions, fluorescent probes for the detection of Co^{2+} in environmental water samples with a limit of detection of 0.37 μmol/L (based on 3 s/k) were reported [70]. Graphene quantum dots (GQDs) were synthesized and incorporated with polyethylene dioxythiophene:poly(4-styrenesulfonate) (PEDOT:PSS) and carbon nanotubes (CNT) to form a composite that can be used as humidity sensors [71].

The creation of innovative nanoribbons and nanobelts using zigzag-hexagonal graphene quantum dots (ZHG) demonstrates distinct electrical, optical, and sensing characteristics, rendering them auspicious options for volatile organic compound (VOC) sensing applications, including trichloroethylene and chloromethane. The sensitivity and selectivity of the nanobelts toward gas molecules are enhanced by their structure, which has a reduced energy gap than that of planar single quantum dots or extended nanoribbons with the interacting electronic states localized at the zigzag edges. VOCs are effectively adsorbed with the help of moderate adsorption energy and charge transfer [72].

2.3.4 Ratiometric fluorescence and colorimetric dual-mode sensors

A ratiometric fluorescence and colorimetric probe is a sophisticated sensing instrument used to identify and measure specific analytes by observing alterations in fluorescence intensity and visible color. It uses two distinct fluorescence signals to react differentially to the presence of a substance, enabling more precise measurements and mitigating environmental fluctuations [73].

Quantum dots (QDs) have been used in the creation of innovative ratiometric fluorescent probes for the precise detection of Cu^{2+} [74], chlorothalonil (CHL) [75], and mercury ions in environmental water samples [76].

2.3.5 Quantum dots in food quality sensing

Food quality and safety are major public health concerns worldwide. Fluorescence detection methods, such as quantum dot (QD)-based fluorescent nanosensors, are gaining popularity due to their broad detection range, high sensitivity, and rapid detection [77]. Different QDs were utilized to detect volatile components, heavy metal ions, food additives, pesticide residues, mycotoxins, foodborne pathogens, humidity, and temperature [78].

Quantum dot-polymer nanocomposites represent notable progress in food packaging, providing immediate monitoring, detection of foodborne pathogens, assessment of food quality, improvement of food safety, and extension of shelf life. These sensors utilize QDs that are incorporated within polymer matrices to monitor temperature, humidity, oxygen levels, and chemical contaminants [79]. Food safety and quality are enhanced by CdS-QDs and chitosan, carboxymethylcellulose, and pectin polymer matrices [80].

3. Non-photonic QDs applications

3.1 QDS in magnetic resonance imaging (MRI) medical imaging

The variable magnetic characteristics of QDs allow customization for specific imaging demands, which is why QDs are now being investigated for use in magnetic resonance imaging (MRI) as contrast agents. It is possible to obtain QDs emitting in the (650–950 nm) and (1000–1400 nm) near-infrared (NIR) windows, by selectively controlling the synthetic methodologies. This allows for superior imaging properties [81].

3.2 QDS in positron emission tomography (PET)

Radiolabeled QDs are important in molecular imaging, notably positron emission tomography (PET), which is used to investigate biological processes in vivo. PET is used to diagnose diseases, track illness progression, and aid in medication discovery in both preclinical and clinical settings. QDs are adaptable radiolabeling probes that broaden the range of synthons acceptable for radiolabeling and enable the creation of PET tracers. These PET tracers have applications in a variety of clinical disciplines, including oncology, cardiology, neurology, and drug discovery, where they improve target engagement studies and enhance drug development efforts. Radiochemical innovations, such as labeling tactics and bioconjugation processes, have expanded the range of functional groups that can be radiolabeled, resulting in the development of novel radiopharmaceuticals. Despite the difficulties with biocompatibility, toxicity, clearance, and stability, QDs in PET imaging have revolutionized the industry, increasing patient care and therapeutic results in the era of precision medicine [82].

4. QDs size-based applications

4.1 Quantum dots (QDs) in drug delivery

Quantum confinement effect and size tunability enable precise targeting and imaging capabilities, offering a new horizon for therapeutic interventions with enhanced efficacy and reduced side effects [83].

A novel method for high-throughput and multiplexed single nucleotide polymorphisms (SNP) genotyping for the purpose of using the Qbeads framework applies QDs to encode microspheres utilized as a stage for multiplexed examines [84]. Qbeads is defined as a family of reagents that enables capture of specific proteins on QDs. This enables multiplexed quantitation of cytokines, adhesion molecules, enzymes, growth factors receptors and more.

On the other hand, QDs are candidates in drug delivery. Specifically, the approach of labeling a conventional drug carrier with QDs, which serve as photostable fluorescent reporters. Is rewritten as: On the other hand, QDs are candidates in drug delivery. One approach is labeling a conventional drug carrier with QDs, which serve as photostable fluorescent reporters. This technique is considered an innovative approach to developing biocompatible nano formulations that can target and treat human diseases by using functionalized nanoparticles to carry medications to specific tissues or organs. Additionally, the passage mentions that QDs have been utilized as drug carriers, both in organic and inorganic forms, and there has been significant progress in the field of ODN (oligonucleotide) and siRNA (small interfering RNA) delivery, potentially even involving microorganisms and viruses [85].

Carbon quantum dots (CQDs) have surfaced as an innovative technology that exhibits great potential for implementation in drug delivery systems. They are great candidates for targeted drug delivery because they are biocompatible, have a surface chemistry that can be changed, and can carry therapeutic agents. Coupled CQDs with gold nanoparticles form an assembly, which was then conjugated with PEI-pDNA for delivering DNA to cells [86]. Functionalized CQDs have the capability to encapsulate and transport a wide range of medications, such as gene therapies, antibiotics, and anticancer agents. CQDs functionalized gold nanorods for the delivery of doxorubicin in a multi-modal fashion, including drug delivery, photothermal therapy, and bioimaging using the same platform [87]. In addition, their fluorescence characteristics facilitate the monitoring of drug distribution in the body in real time, which enables accurate administration and reduces the occurrence of adverse effects. Drug delivery systems utilizing CQDs have the capacity to fundamentally transform the medical field through the enhancement of treatment efficacy and safety across a broad spectrum of ailments. The antibiotic ciprofloxacin was attached to CQDs with bright green fluorescence allowing bioimaging and providing an efficient new nanocarrier for controlled drug release with high antimicrobial activity under physiological conditions [88].

4.2 Quantum dots (QDs) in blood-brain barrier (BBB) crossing

The BBB serves as a highly selective semipermeable border that separates the circulating blood from the brain and extracellular fluid in the central nervous system. It restricts the passage of most chemical substances, ensuring the brain's protection from potentially harmful agents while also posing a significant challenge for the delivery of therapeutic agents [89]. The significance of QDs in BBB lies in its potential to modify or enhance the permeability of the BBB, enabling the delivery of drugs that would otherwise be unable to reach the brain tissue. By integrating QDs into the design of drug delivery systems, the efficiency of treatments for neurological disorders can be improved, where the ability to bypass or temporarily open the BBB could significantly enhance therapeutic outcomes. This approach holds promise for advancing the treatment of a wide range of conditions, including neurodegenerative diseases, brain tumors, and psychiatric disorders, by facilitating targeted drug delivery directly to the site of pathology [90]. Several studies have investigated QDs for efficient blood-brain barrier (BBB) penetration and brain-targeted drug delivery. The successful use of asparagine-glycine-arginine peptide NGR-PEG-CdSe/ZnS QDs is suitable for imaging malignant brain tumors, where QDs could cross the BBB and target CD13-overexpressing glioma [91], CM-QDs-PEG-PAEA-DDA was used for cellular brain imaging [92]. CdSe@ZnS/ZnS core@multiple shell QDs were used as a fluorescence imaging probe with a size of 12.7 nm (CM-QDs-PEG-PAEA-DDA). The probe has high solubility, stability, and biocompatibility and enters the brain within 1 hour [93]. PEGylated Ag_2S QDs exhibiting robust near-infrared (NIR) fluorescence were successfully produced and employed to dynamically visualize intracerebral hemorrhage and increased blood-brain barrier (BBB) permeability induced by hyperglycinemia, thereby facilitating the enhanced accumulation of QDs within the brain [94]. Polyethylene glycol (PEG) was utilized as a coating for MoS_2 quantum dots (QDs) to enable the attachment of a mitochondria-targeted ligand, resulting in effective penetration through the blood-brain barrier (BBB) [95]. Water-soluble zinc oxide quantum dots (ZnO QDs) were employed as carriers to deliver pDNA into the brain for Alzheimer's disease (AD) therapy. Also, glutathione (GSH) was firmly attached to the nanoparticles (NPs) to help them pass through the blood-brain barrier (BBB) more easily through a process called receptor-mediated transcytosis (RMT) [96, 97].

Both carbon quantum dots (CQDs) and graphene quantum dots (GQDs) are promising candidates capable of penetrating the BBB. Both types of QDs show promising progress in drug delivery for the treatment of CNS diseases [98, 99].

Additionally, the use of QDs in blood-brain barrier research aids in the development of diagnostic techniques and imaging modalities. QDs' unique optical properties, including intense and stable fluorescence, enable precise visualization of BBB permeability and integrity. This imaging capability is particularly valuable in understanding the dynamics of diseases that affect the BBB, such as brain tumors or inflammatory conditions [97]. QD-based imaging provides researchers with a non-invasive and sensitive approach to detect subtle changes in the BBB structure and function, facilitating early diagnosis and monitoring of neurological disorders. The integration of QD technology in blood-brain barrier studies contributes to the advancement of both basic neuroscience research and the development of innovative diagnostic and therapeutic strategies for neurological conditions (**Figure 4**).

4.3 Quantum dots in electrochemical sensing

QDs have revolutionized electrochemical sensing, particularly in biosensing. Graphene quantum dots (GQDs) are a promising immobilizing agent for cardiac biomarkers, proving useful in early diagnosis of myocardial infarction. Their size-dependent properties enable precise detection mechanisms, and their high surface-volume

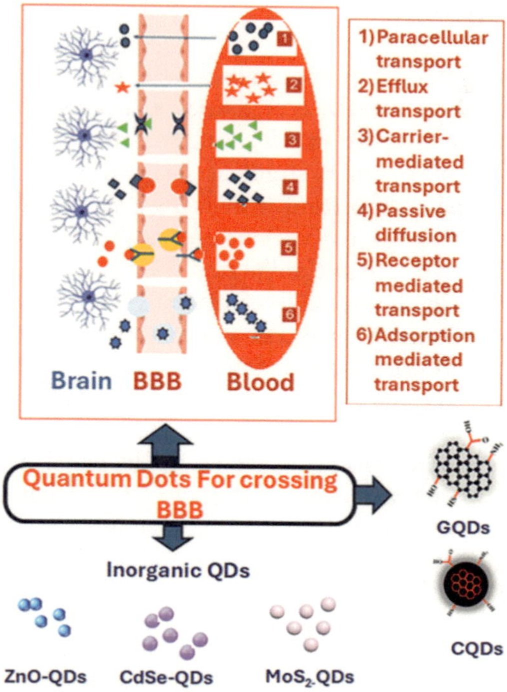

Figure 4.
The mechanisms of external materials crossing BBB.

ratio, biocompatibility, and surface modification capabilities make them valuable in immunological electrochemical sensors [100].

QDs have proven their versatility in various sensing applications, such as multiplex immunosensors using PbS, ZnS, and CdS QDs, and cytochrome C protein detection on gold surfaces. QDs have also been used in electrochemical sensing, generating measurable currents through electrochemiluminescence. These examples demonstrate the broad applicability and innovative use of QDs in electrochemical sensing, paving the way for advanced diagnostics and analytical techniques [5].

Two-dimensional quantum dots (2D-QDs) are a versatile material with unique properties, including large surface areas, abundant active sites, excellent electrical properties, good aqueous dispersibility, and ease of functionalization. They are used in electrochemical biosensing for various types of sensors, such as DNA, immunological, enzyme, and apt sensors. 2D-QDs such as graphene, black phosphorus, nitrides, transition metal dichalcogenides (TMDCs), transition metal oxides (TMOs), and MXenes offer a broad spectrum of characteristics due to quantum confinement effects and edge effects, which are beneficial for sensing applications. Their advantages include easy immobilization of biomolecules, biocompatibility, and high multifunctionality. They also serve as co-reactants and emitters in electrochemiluminescence biosensors. Their high-surface activity, strong adsorption capacity, and improved electron transfer efficiency make them ideal for electrode modification. 2D-QDs are also used in chemical sensing, bioimaging, solar cells, and electromagnetic wave absorption, offering high sensitivity, selectivity, and low detection limits for various analytes [101].

4.4 QDs as photocatalysts

QDs have carved out a place in photocatalysis, opening new opportunities for energy conversion and environmental remediation solutions. The QDs' high surface-to-volume ratio offers several active sites for catalytic processes, hence augmenting their activity. Furthermore, the ability to adjust the band gap of quantum dots (QDs) by modifying their size and composition enables the customized creation of photocatalysts that can efficiently engage in various processes, such as water splitting, CO_2 reduction, and degradation of organic pollutants [102].

The utilization of QDs in photocatalysis and catalysis encounters obstacles, such as the vulnerability to agglomeration and photo-corrosion, which can hinder their effectiveness. Recent progress has been centered around tackling these problems by employing techniques such as surface modification [103], heterostructure with other materials, and the integration of protective layers [102]. Surface functionalization of quantum dots significantly increases the surface dipole, which affects the valence and conduction and improves the catalytic performance [104]. Quaternary photocatalyst quantum dots exhibit significant visible-light photoactivity and quick charge migration, revealing QDs' ability to increase the photocatalytic destruction of pollutants under visible light [105].

Green QDs can be used as photocatalysts or coupled with other semiconductors to make heterojunctions, which improve charge separation and reduce electron-hole recombination. This improves the efficiency of processes such as organic pollutant degradation, CO_2 conversion to useful compounds, and water splitting for hydrogen production. Green QDs adhere to the principles of green chemistry, providing a low-toxicity and ecologically friendly alternative to standard photocatalysts. *Citrus grandis*-derived CQDs have been explored for metal-free photocatalysis for pollution removal under sunlight, demonstrating an eco-friendly approach [106]. Palm powder-derived CQDs co-doped with sulfur/chlorine show promising applications in visible-light photocatalysis [107].

Different applications of QDS in photocatalysis include:

4.4.1 Photocatalytic CO2 reduction

Photocatalytic CO_2 reduction aims to convert CO_2 into valuable organic fuels, using quantum confinement effects as a low-cost and efficient catalyst [108]. An innovative approach involves encapsulating CuO QDs within a MIL-125 (Ti) MOF and integrating $LiFePO_4$ to promote photoreduction [109]. This reduces bandgap, enhancing electron transfer and hole transfer, preventing recombination. Supporting CdS QDs on a 3D-ordered macroporous carbon skeleton increases CO_2 adsorption and provides active sites for CO_2 reduction [110].

4.4.2 Photocatalytic H2 production

Quantum dots (QDs) are crucial for hydrogen production through photocatalytic water splitting, a process essential for sustainable energy generation. Their tuning capability maximizes solar energy conversion efficiency, a critical factor in photocatalytic hydrogen production. Photocatalytic hydrogen evolution was improved using a p-n heterojunction of n-type NiP_2 quantum dots and p-type Cu_3P nanoparticles from NiAl-LDH as a powerful photocatalyst [111].

4.4.3 Photocatalytic environmental remediation

QDs can be optimized for specific remediation tasks, such as dye degradation and heavy metal ion removal from water. Anatase/rutile QDs on thin slices of g-C_3N_4 were employed as a photocatalyst to degrade oxytetracycline in salted seawater [112]. Surface modification can enhance selectivity and efficiency. However, practical application poses challenges, including potential toxicity and environmental impact [113].

5. Conclusion and prospects

QDs are entities of great unique optical, electronic, and chemical properties that are not observed in other materials. These unique properties are attractive for a variety of scientific and commercial applications. Photonic applications include, for example, quantum dot lasers, light-emitting diodes (LED), displays, solar cells, and biosensors. Non-photonic applications include magnetic QDs, magnetic resonance imaging (MRI), positron emission tomography (PET), radiolabeling, drug delivery, blood-brain barrier (BBB) crossing, electrochemical sensing, photocatalysis, environment remediation, photocatalytic CO_2 reduction, green H_2 production. Interesting potential QD applications in future perspectives encounter crucial industries such as displays, solar cells, wastewater treatment, quantum computers, and biomedical applications. However, the major drawback of traditional semiconductor QDs made of heavy metal ions is their toxicity. Core-shell structure modification of QDs applying biocompatible ligands or polymers is one way to minimize toxicity. Heavy metal-free QDs formulations using non-toxic and eco-friendly starting materials include carbon-based, biomolecules-based, silicon-based, and ternary I-III-VI QDs are more promising alternatives.

Author details

El-Zeiny M. Ebeid* and Ehab A. Okba
Faculty of Science, Chemistry Department, Tanta University, Tanta, Egypt

*Address all correspondence to: elzeiny.ebeid@science.tanta.edu.eg

References

[1] Kim D, Yoo S. Aptamer-conjugated quantum dot optical biosensors: Strategies and applications. Chem. 2021;**9**:318. DOI: 10.3390/chemosensors9110318

[2] Azimirad R, Najafi V, Safa S. Tunable emission of the laser reduced graphene quantum dots be used as wavelength shifter layers in silicon-based UV detectors. Optics and Laser Technology. 2024;**174**:110591. DOI: 10.1016/j.optlastec.2024.110591

[3] Heng ZW, Chong WC, Pang YL, Koo CH. An overview of the recent advances of carbon quantum dots/metal oxides in the application of heterogeneous photocatalysis in photodegradation of pollutants towards visible-light and solar energy exploitation. Journal of Environmental Chemical Engineering. 2021;**9**:105199. DOI: 10.1016/j.jece.2021.105199

[4] Shilpa G, Kumar PM, Kumar DK, Deepthi P, Sadhu V, Sukhdev A, et al. Recent advances in the development of high efficiency quantum dot sensitized solar cells (QDSSCs): A review. Materials Science for Energy Technologies. 2023;**6**:533-546. DOI: 10.1016/j.mset.2023.05.001

[5] Carvalho JPR, De-Simone SG. Electro Sensors Based on Quantum Dots and Their Applications in Diagnostic Medicine. London, United Kingdom: IntechOpen; 2023

[6] Dessale M, Mengistu G, Mengist HM. Nanotechnology: A promising approach for cancer diagnosis, therapeutics and theragnosis. International Journal of Nanomedicine. 2022;**17**:3735. DOI: 10.2147%2FIJN.S378074

[7] Badıllı U, Mollarasouli F, Bakirhan NK, Ozkan Y, Ozkan SA. Role of quantum dots in pharmaceutical and biomedical analysis, and its application in drug delivery. TrAC, Trends in Analytical Chemistry. 2020;**131**:116013. DOI: 10.1016/j.trac.2020.116013

[8] Facure MH, Schneider R, Lima JB, Mercante LA, Correa DS. Graphene quantum dots-based nanocomposites applied in electrochemical sensors: A recent survey. Electrochemistry. 2021;**2**:490-519. DOI: 10.3390/electrochem2030032

[9] Katwal R. A review: Properties and diverse applications of smart magnetic quantum dots. Nano-Structures & Nano-Objects. 2023;**35**:101001. DOI: 10.1016/j.nanoso.2023.101001

[10] Khan SU, Eren GO, Atac N, Onal A, Qureshi MH, Cooper FK, et al. Antibacterial type-II InP/ZnO quantum dots via multimodal reactive oxygen species. Chemical Engineering Journal. 2024;**480**:148140. DOI: 10.1016/j.cej.2023.148140

[11] Rajendiran K, Zhao Z, Pei D-S, Fu A. Antimicrobial activity and mechanism of functionalized quantum dots. Polymers. 2019;**11**:1670. DOI: 10.3390/polym11101670

[12] Levy M, Chowdhury PP, Nagpal P. Quantum dot therapeutics: A new class of radical therapies. Journal of Biological Engineering. 2019;**13**:1-12. DOI: 10.1186/s13036-019-0173-4

[13] Sun H. Analysis and demonstration of the quantum dot lasers. Highlights in Science, Engineering and Technology. 2023;**38**:1000-1005. DOI: 10.54097/hset.v38i.5988

[14] Norman JC, Mirin RP, Bowers JE. Quantum dot lasers—History and future prospects. Journal of Vacuum Science and Technology A. 2021;**39**:050802-1 to 050802-23. DOI: 10.1116/6.0000768

[15] Potter KS, Simmons JH. Chapter 7—Optical gain and lasers. In: Potter KS, Simmons JH, editors. Optical Materials. 2nd ed. Netherlands: Elsevier; 2021. pp. 369-428. DOI: 10.1016/B978-0-12-818642-8.00007-7

[16] Yao Z, Jiang C, Wang X, Chen H, Wang H, Qin L, et al. Recent developments of quantum dot materials for high speed and ultrafast lasers. Nanomaterials. 2022;**12**:1058

[17] Wu Y, Jiang L, Asryan LV. Output power of a quantum dot laser: Effects of excited states. Journal of Applied Physics. 2015;**118**:193105-1 to 193105-6. DOI: 10.1063/1.4935296

[18] Plappert S, Liebner F. Cellulose-based photoluminescent nanocomposites. In: Lignocellulosics. Netherlands: Elsevier; 2020. pp. 117-170. DOI: 10.1016/B978-0-12-8040775.00010-5

[19] Auzel F. Upconversion and anti-stokes processes with f and d ions in solids. Chemical Reviews. 2004;**104**:139-174. DOI: 10.1021/cr020357g

[20] Dutta R, Barik P. Upconversion and downconversion quantum dots for biomedical and therapeutic applications. In: Application of Quantum Dots in Biology and Medicine: Recent Advances. Germany: Springer; 2022. pp. 229-263. DOI: 10.1007/978-981-19-3144-4

[21] Agranovich VM, Maradudin A. Spectroscopy of Solids Containing Rare Earth Ions. Netherlands: North-Holland Publishing Company; 1987

[22] Singh P, Singh P, Singh S. Photon upconversion spectroscopy. In: Modern Techniques of Spectroscopy: Basics, Instrumentation, and Applications. Germany: Springer; 2021. pp. 389-407. DOI: 10.1007/978-981-33-6084-6

[23] Chivian JS, Case W, Eden D. The photon avalanche: A new phenomenon in Pr3+-based infrared quantum counters. Applied Physics Letters. 1979;**35**:124-125. DOI: 10.1063/1.91044

[24] Kueny A, Case WE, Koch ME. Nonlinear-optical absorption through photon avalanche. Journal of the Optical Society of America B. 1989;**6**:639-642. DOI: 10.1364/JOSAB.6.000639

[25] Nguyen P-D, Son SJ, Min J. Upconversion nanoparticles in bioassays, optical imaging and therapy. Journal of Nanoscience and Nanotechnology. 2014;**14**:157-174. DOI: 10.1166/jnn.2014.8894

[26] Chen G, Qiu H, Prasad PN, Chen X. Upconversion nanoparticles: Design, nanochemistry, and applications in theranostics. Chemical Reviews. 2014;**114**:5161-5214. DOI: 10.1021/cr400425h

[27] Zou Q, Huang P, Zheng W, You W, Li R, Tu D, et al. Cooperative and non-cooperative sensitization upconversion in lanthanide-doped LiYbF4 nanoparticles. Nanoscale. 2017;**9**:6521-6528. DOI: 10.1039/C7NR02124K

[28] Cevallos-Toledo RB, Bellezza D, Ferrera-González J, Giussani A, Ortí E, González-Béjar M, et al. Cooperative sensitization upconversion in ytterbium (III)-based eosin Lake pigments. ChemPhotoChem. 2023;**7**:e202300156. DOI: 10.1002/cptc.202300156

[29] Zhou B, Shi B, Jin D, Liu X. Controlling upconversion nanocrystals for emerging applications. Nature

Nanotechnology. 2015;**10**:924-936. DOI: 10.5772/65118

[30] Liu G. Advances in the theoretical understanding of photon upconversion in rare-earth activated nanophosphors. Chemical Society Reviews. 2015;**44**: 1635-1652. DOI: 10.1039/C4CS00187G

[31] Wang K, Cline RP, Schwan J, Strain JM, Roberts ST, Mangolini L, et al. Efficient photon upconversion enabled by strong coupling between silicon quantum dots and anthracene. Nature Chemistry. 2023;**15**:1172-1178. DOI: 10.1038/s41557-023-01225-x

[32] Huang Z, Tung C-H, Wu L-Z. Quantum dot-sensitized triplet–triplet annihilation photon upconversion for solar energy conversion and beyond. Accounts of Materials Research. 2024;**5**:136-145. DOI: 10.1021/accountsmr.3c00186

[33] Wang F, Deng R, Wang J, Wang Q, Han Y, Zhu H, et al. Tuning upconversion through energy migration in core–shell nanoparticles. Nature Materials. 2011;**10**:968-973. DOI: 10.1038/nmat3149

[34] Mahata MK, Hofsäss HC, Vetter U. Photon-upconverting materials: Advances and prospects for various emerging applications. In: Luminescence—An Outlook on the Phenomena and Their Applications. Cambridge, United Kingdom: Royal Society of Chemistry (RSC); 2016. pp. 109-131

[35] Sun L-D, Dong H, Zhang P-Z, Yan C-H. Upconversion of rare earth nanomaterials. Annual Review of Physical Chemistry. 2015;**66**:619-642. DOI: 10.1146/annurev-physchem-040214-121344

[36] Chen X, Peng D, Ju Q, Wang F. Photon upconversion in core–shell nanoparticles. Chemical Society Reviews. 2015;**44**:1318-1330. DOI: 10.1039/C4CS00151F

[37] Wegh RT, Donker H, Oskam KD, Meijerink A. Visible quantum cutting in LiGdF4: Eu3+ through downconversion. Science. 1999;**283**:663-666. DOI: 10.1126/science.283.5402.663

[38] Zou Z, Feng L, Cao C, Zhang J, Wang Y. Near-infrared quantum cutting long persistent luminescence. Scientific Reports. 2016;**6**:24884. DOI: 10.1038/srep24884

[39] Ayoub I, Mushtaq U, Swart HC, Kumar V. 14—Metal oxides in quantum-dot-based LEDs and their applications. In: Kumar V, Sharma V, Swart HC, Das S, editors. Metal Oxides for Next-Generation Optoelectronic, Photonic, and Photovoltaic Applications. Amsterdam, Netherlands: Elsevier; 2024. pp. 409-442. DOI: 10.1016/B978-0-323-99143-8.00010-9

[40] Zhang Z, Shen L, Zhang H, Ding L, Shao G, Liang X, et al. Novel red-emitting CsPb1−xTixI3 perovskite QDs@glasses with ambient stability for high efficiency white LEDs and plant growth LEDs. Chemical Engineering Journal. 2019;**378**:122125. DOI: 10.1016/j.cej.2019.122125

[41] Wang HC, Bao Z, Tsai HY, Tang AC, Liu RS. Perovskite quantum dots and their application in light-emitting diodes. Small. 2018;**14**:1702433. DOI: 10.1002/smll.201702433

[42] Mei X, Jia D, Chen J, Zheng S, Zhang X. Approaching high-performance light-emitting devices upon perovskite quantum dots: Advances and prospects. Nano Today. 2022;**43**:101449. DOI: 10.1016/j.nantod.2022.101449

[43] Wang HC, Wang W, Tang AC, Tsai HY, Bao Z, Ihara T, et al.

High-performance CsPb1−xSnxBr3 perovskite quantum dots for light-emitting diodes. Angewandte Chemie. 2017;**129**:13838-13842. DOI: 10.1002/anie.201706860

[44] Mei X, He K, Zhuang R, Yu M, Hua Y, Zhang X. Stabilizing dynamic surface of highly luminescent perovskite quantum dots for light-emitting diodes. Chemical Engineering Journal. 2023;**453**:139909. DOI: 10.1016/j.cej.2022.139909

[45] Armăşelu A. Recent developments in applications of quantum-dot based light-emitting diodes. In: Quantum-Dot Based Light-emitting Diodes. London, United Kingdom: IntechOpen; 2017. DOI: 10.5772/intechopen.69177

[46] Kwak HJ, Kiguye C, Gong M, Park JH, Kim G-H, Kim JY. Enhanced performance of inverted perovskite quantum dot light-emitting diode using electron suppression layer and surface morphology control. Materials. 2023;**16**:7171. DOI: 10.3390/ma16227171

[47] Chae Y-B, Kim S-Y, Choi H-D, Moon D-G, Lee K-H, Kim C-K. Enhancing efficiency in inverted quantum dot light-emitting diodes through arginine-modified ZnO nanoparticle electron injection layer. Nanomaterials. 2024;**14**:266. DOI: 10.3390/nano14030266

[48] He Y, Rao H, Wang J, Wu Y, Han C, Yan C, et al. Perovskite quantum dots modulating upconversion nanomaterials for cancer early detections. Cancer Nanotechnology. 2023;**14**:52. DOI: 10.1186/s12645-023-00206-6

[49] Lai R, Wu K. Red-to-blue photon upconversion based on a triplet energy transfer process not retarded but enabled by shell-coated quantum dots. The Journal of Chemical Physics. 2020;**153**. DOI: 10.1063/5.0023052

[50] Lu H, Huang Z, Martinez MS, Johnson JC, Luther JM, Beard MC. Transforming energy using quantum dots. Energy & Environmental Science. 2020;**13**:1347-1376. DOI: 10.1039/C9EE03930A

[51] Roy P, Virmani M, Pillai PP. Blue-emitting InP quantum dots participate in an efficient resonance energy transfer process in water. Chemical Science. 2023;**14**:5167-5176

[52] Ramezani Z, Thompson M. How Functionalization Affects the Detection Ability of Quantum Dots. Cambridge, United Kingdom: The Royal Society of Chemistry; 2023. DOI: 10.1039/9781839169564-00037

[53] Mohkam M, Sadraeian M, Lauto A, Gholami A, Nabavizadeh SH, Esmaeilzadeh H, et al. Exploring the potential and safety of quantum dots in allergy diagnostics. Microsystems & Nanoengineering. 2023;**9**:145. DOI: 0.1038/s41378-023-00608-x

[54] Yu G-T, Luo M-Y, Li H, Chen S, Huang B, Sun Z-J, et al. Molecular targeting nanoprobes with non-overlap emission in the second near-infrared window for in vivo two-color colocalization of immune cells. ACS Nano. 2019;**13**:12830-12839. DOI: 10.1021/acsnano.9b05038

[55] Zhao P, Xu Q, Tao J, Jin Z, Pan Y, Yu C, et al. Near infrared quantum dots in biomedical applications: Current status and future perspective. Wiley Interdisciplinary Reviews: Nanomedicine and Nanobiotechnology. 2018;**10**:e1483

[56] Resch-Genger U, Grabolle M, Cavaliere-Jaricot S, Nitschke R, Nann T. Quantum dots versus organic dyes as fluorescent labels. Nature Methods. 2008;**5**:763-775. DOI: 10.1002/wnan.1483

[57] Pons T, Bouccara S, Loriette V, Lequeux N, Pezet S, Fragola A. In vivo imaging of single tumor cells in fast-flowing bloodstream using near-infrared quantum dots and time-gated imaging. ACS Nano. 2019;**13**:3125-3131. DOI: 10.1021/acsnano.8b08463

[58] Stasiuk GJ, Tamang S, Imbert D, Poillot C, Giardiello M, Tisseyre C, et al. Cell-permeable Ln (III) chelate-functionalized InP quantum dots as multimodal imaging agents. ACS Nano. 2011;**5**:8193-8201. DOI: 10.1021/nn202839w

[59] Khaledian S, Abdoli M, Fatahian R, Zahabi SS. Quantum dots in cancer cell imaging. In: Thirumalai J, editor. Quantum Dots-Recent Advances, New Perspectives and Contemporary Applications. London, UK, London, UK: IntechOpen; 2023. DOI: 10.5772/intechopen.102101

[60] Shi W, Wang Q, Long Y, Cheng Z, Chen S, Zheng H, et al. Carbon nanodots as peroxidase mimetics and their applications to glucose detection. Chemical Communications. 2011;**47**:6695-6697. DOI: 10.1039/c1cc11943e

[61] Zhu A, Qu Q, Shao X, Kong B, Tian Y. Carbon-dot-based dual-emission nanohybrid produces a ratiometric fluorescent sensor for in vivo imaging of cellular copper ions. Angewandte Chemie, International Edition. 2012;**51**:7185-7189. DOI: 10.1002/anie.201109089

[62] Zhao HX, Liu LQ, De Liu Z, Wang Y, Zhao XJ, Huang CZ. Highly selective detection of phosphate in very complicated matrixes with an off–on fluorescent probe of europium-adjusted carbon dots. Chemical Communications. 2011;**47**:2604-2606

[63] Zhu S, Meng Q, Wang L, Zhang J, Song Y, Jin H, et al. Highly photoluminescent carbon dots for multicolor patterning, sensors, and bioimaging. Angewandte Chemie, International Edition. 2013;**52**:3953-3957. DOI: 10.1002/anie.201300519

[64] Wei W, Xu C, Ren J, Xu B, Qu X. Sensing metal ions with ion selectivity of a crown ether and fluorescence resonance energy transfer between carbon dots and graphene. Chemical Communications. 2012;**48**:1284-1286. DOI: 10.1039/c2cc16481g

[65] Shi W, Li X, Ma H. A tunable ratiometric pH sensor based on carbon nanodots for the quantitative measurement of the intracellular pH of whole cells. Angewandte Chemie, International Edition. 2012;**51**:6432-6435. DOI: 10.1002/anie.201202533

[66] Li H, Zhang Y, Wang L, Tian J, Sun X. Nucleic acid detection using carbon nanoparticles as a fluorescent sensing platform. Chemical Communications. 2011;**47**:961-963

[67] Mousavi SM, Hashemi SA, Yari Kalashgrani M, Omidifar N, Lai CW, Vijayakameswara Rao N, et al. The pivotal role of quantum dots-based biomarkers integrated with ultra-sensitive probes for multiplex detection of human viral infections. Pharmaceuticals. 2022;**15**:880. DOI: 10.3390/ph15070880

[68] Klestova Z, Voronina A, Yushchenko AY, Vatlitsova O, Dorozinsky G, Ushenin YV, et al. Aspects of "antigen–antibody" interaction of chicken infectious bronchitis virus determined by surface plasmon resonance. Spectrochimica Acta. Part A, Molecular and Biomolecular Spectroscopy. 2022;**264**:120236. DOI: 10.1016/j.saa.2021.120236

[69] Park S, Cho E, Chueng S-TD, Yoon J-S, Lee T, Lee J-H. Aptameric fluorescent biosensors for liver cancer diagnosis. Biosensors. 2023;**13**:617. DOI: 10.3390/bios13060617

[70] Sullam EM, Adam KM, Liu J, Chen H, Xiao J. Silicon quantum dots-based fluorescent sensor for the detection of cobalt with high sensitivity and selectivity. Chinese Chemical Letters. 2024;**35**:108476

[71] Long LM, Dinh NN, Trung TQ. Synthesis and characterization of polymeric graphene quantum dots based nanocomposites for humidity sensing. Journal of Nanomaterials. 2016;**2016**: 1-12. DOI: 10.1155/2016/5849018

[72] Abdelsalam H, Abd-Elkader OH, Sakr MAS, Teleb NH, Osman W, Zhilong W, et al. Novel nanobelts constructed from hexagonal graphene quantum dots: Electronic, optical, and sensing properties. Results in Physics. 2024;**57**:107412. DOI: 10.1016/j.rinp.2024.107412

[73] Chen X, Zhuang Y, Chen J, Lin J, Chen J, Han Z. Novel ratiometric fluorescence and colorimetric dual-mode biosensors for Cu2+ and biothiols detection based on norepinephrine modified carbon dots. Microchemical Journal. 2022;**181**:107828. DOI: 10.1016/j.microc.2022.107828

[74] Zhang H, Li Y, Lu H, Gan F. A ratiometric fluorescence and colorimetric dual-mode sensing platform based on sulfur quantum dots and carbon quantum dots for selective detection of Cu2+. Analytical and Bioanalytical Chemistry. 2022;**414**:2471-2480

[75] Sheng E, Lu Y, Tan Y, Xiao Y, Li Z, Dai Z. Ratiometric fluorescent quantum dot-based biosensor for chlorothalonil detection via an inner-filter effect. Analytical Chemistry. 2020;**92**:4364-4370. DOI: 10.1021/acs.analchem.9b05199

[76] Liu Y, Su X, Liu H, Zhu G, Ge G, Wang Y, et al. Construction of eco-friendly dual carbon dots ratiometric fluorescence probe for highly selective and efficient sensing mercury ion. Journal of Environmental Sciences. 2024;**148**:1-12. DOI: 10.1016/j.jes.2024.01.013

[77] Das TK, Ganguly S. Revolutionizing food safety with quantum dot–polymer nanocomposites: From monitoring to sensing applications. Food. 2023;**12**:2195. DOI: 0.3390/foods 12112195

[78] Jia Z, Shi C, Yang X, Zhang J, Sun X, Guo Y, et al. QD-based fluorescent nanosensors: Production methods, optoelectronic properties, and recent food applications. Comprehensive Reviews in Food Science and Food Safety. 2023;**22**:4644-4669. DOI: 10.1111/1541-4337.13236

[79] Savelli CJ, Mateus C. Exploring the international food safety authorities network as a community of practice: Results from a global survey of network members. Journal of Food Protection. 2021;**84**:262-274. DOI: 10.4315/jfp-20-313

[80] Méndez PA, Ramírez K, Lucero A, Rodríguez J, López B. Fluorescent nanocomposites based on cadmium sulfide quantum dots and cationic/anionic polymer matrices: Characterization and application as sensor for visual detection of glyphosate. In: Anionic Polymer Matrices: Characterization and Application as Sensor for Visual Detection of Glyphosate. Germany: Springer; 2023

[81] Gil HM, Price TW, Chelani K, Bouillard J-SG, Calaminus SD, Stasiuk GJ. NIR-quantum dots in

biomedical imaging and their future. iScience. 2021;**24**:102189-1 to 102189-10. DOI: 10.1016/j.isci.2021.102189

[82] Rong J, Haider A, Jeppesen TE, Josephson L, Liang SH. Radiochemistry for positron emission tomography. Nature Communications. 2023;**14**:3257. DOI: 10.1038/s41467-023-36377-4

[83] Bhurbhure O, Ghormade V, Katekar V, Sangule D, Dhage S, Boralkar V, et al. Quantum dots: A new hope for the pharmaceutical field. Journal of Drug Delivery and Therapeutics. 2022;**12**:236-244. DOI: 10.22270/jddt.v12i3-s.5373

[84] Tabatabaei Mirakabad FS, Nejati-KoshkiK,AkbarzadehA,YamchiMR, Milani M, Zarghami N, et al. PLGA-based nanoparticles as cancer drug delivery systems. Asian Pacific Journal of Cancer Prevention. 2014;**15**:517-535. DOI: 10.7314/apjcp.2014.15.2.517

[85] Michalet X, Pinaud FF, Bentolila LA, Tsay JM, Doose S, Li JJ, et al. Quantum dots for live cells, in vivo imaging, and diagnostics. Science. 2005;**307**:538-544. DOI: 10.1126/science.1104274

[86] Kim J, Park J, Kim H, Singha K, Kim WJ. Transfection and intracellular trafficking properties of carbon dot-gold nanoparticle molecular assembly conjugated with PEI-pDNA. Biomaterials. 2013;**34**:7168-7180. DOI: 10.1016/j.biomaterials.2013.05.072

[87] Pandey S, Thakur M, Mewada A, Anjarlekar D, Mishra N, Sharon M. Carbon dots functionalized gold nanorod mediated delivery of doxorubicin: Tri-functional nano-worms for drug delivery, photothermal therapy and bioimaging. Journal of Materials Chemistry B. 2013;**1**:4972-4982. DOI: 10.1039/C3TB20761G

[88] Thakur M, Pandey S, Mewada A, Patil V, Khade M, Goshi E, et al. Antibiotic conjugated fluorescent carbon dots as a theranostic agent for controlled drug release, bioimaging, and enhanced antimicrobial activity. Journal of Drug Delivery. 2014;**2014**:1-9. DOI: 10.1155/2014/282193

[89] Wu D, Chen Q, Chen X, Han F, Chen Z, Wang Y. The blood–brain barrier: Structure, regulation, and drug delivery. Signal Transduction and Targeted Therapy. 2023;**8**:217

[90] Tan Q, Zhao S, Xu T, Wang Q, Zhang M, Yan L, et al. Inorganic nano-drug delivery systems for crossing the blood–brain barrier: Advances and challenges. Coordination Chemistry Reviews. 2023;**494**:215344. DOI: 10.1016/j.ccr.2023.215344

[91] Tang Y, Han S, Liu H, Chen X, Huang L, Li X, et al. The role of surface chemistry in determining in vivo biodistribution and toxicity of CdSe/ZnS core–shell quantum dots. Biomaterials. 2013;**34**:8741-8755. DOI: 10.1016/j.biomaterials.2013.07.087

[92] Huang N, Cheng S, Zhang X, Tian Q, Pi J, Tang J, et al. Efficacy of NGR peptide-modified PEGylated quantum dots for crossing the blood–brain barrier and targeted fluorescence imaging of glioma and tumor vasculature. Nanomedicine. 2017;**13**:83-93. DOI: 10.1016/j.nano.2016.08.029

[93] Yang HY, Fu Y, Jang M-S, Li Y, Yin WP, Ahn TK, et al. CdSe@ ZnS/ZnS quantum dots loaded in polymeric micelles as a pH-triggerable targeting fluorescence imaging probe for detecting cerebral ischemic area. Colloids and Surfaces, B: Biointerfaces. 2017;**155**:497-506. DOI: 10.1016/j.colsurfb.2017.04.054

[94] Parchur AK, Fang Z, Jagtap JM, Sharma G, Hansen C, Shafiee S, et al. NIR-II window tracking of hyperglycemia induced intracerebral hemorrhage in cerebral cavernous malformation deficient mice. Biomaterials Science. 2020;**8**:5133-5144. DOI: 10.1039/D0BM00873G

[95] Ren C, Li D, Zhou Q, Hu X. Mitochondria-targeted TPP-MoS2 with dual enzyme activity provides efficient neuroprotection through M1/M2 microglial polarization in an Alzheimer's disease model. Biomaterials. 2020;**232**:119752. DOI: 10.1016/j.biomaterials.2019.119752

[96] Lin D, Li M, Gao Y, Yin L, Guan Y. Brain-targeted gene delivery of ZnO quantum dots nanoplatform for the treatment of Parkinson disease. Chemical Engineering Journal. 2022;**429**:132210. DOI: 10.1016/j.cej.2021.132210

[97] Chen HJ, Qin Y, Wang ZG, Wang L, Pang DW, Zhao D, et al. An activatable and reversible virus-mimicking NIR-II nanoprobe for monitoring the progression of viral encephalitis. Angewandte Chemie, International Edition. 2022;**61**:e202210285. DOI: 10.1002/anie.202210285

[98] Perini G, Palmieri V, Ciasca G, De Spirito M, Papi M. Unravelling the potential of graphene quantum dots in biomedicine and neuroscience. International Journal of Molecular Sciences. 2020;**21**:3712. DOI: 10.3390/ijms21103712

[99] Molaei MJ. Carbon quantum dots and their biomedical and therapeutic applications: A review. RSC Advances. 2019;**9**:6460-6481. DOI: 10.1039/c8ra08088g

[100] Tabish TA, Hayat H, Abbas A, Narayan RJ. Graphene quantum dots-based electrochemical biosensing platform for early detection of acute myocardial infarction. Biosensors. 2022;**12**:77. DOI: 10.3390/bios12020077

[101] Zhang J, Zhang X, Bi S. Two-dimensional quantum dot-based electrochemical biosensors. Biosensors. 2022;**12**:254. DOI: 10.3390/bios12040254

[102] Sun P, Xing Z, Li Z, Zhou W. Recent advances in quantum dots photocatalysts. Chemical Engineering Journal. 2023;**458**:141399. DOI: 10.1016/j.cej.2023.141399

[103] Liang Q, Zhang C, Xu S, Zhou M, Zhou Y, Li Z. In situ growth of CdS quantum dots on phosphorus-doped carbon nitride hollow tubes as active 0D/1D heterostructures for photocatalytic hydrogen evolution. Journal of Colloid and Interface Science. 2020;**577**:1-11. DOI: 10.1016/j.jcis.2020.05.053

[104] Yang W, Yang Y, Kaledin AL, He S, Jin T, McBride JR, et al. Surface passivation extends single and biexciton lifetimes of InP quantum dots. Chemical Science. 2020;**11**:5779-5789

[105] Li W, Zhao C, Zhang Q. Synthesis of Bi/BiOCl-TiO2-CQDs quaternary photocatalyst with enhanced visible-light photoactivity and fast charge migration. Catalysis Communications. 2018;**107**:74-77. DOI: 10.1016/j.catcom.2017.11.011

[106] Ramar V, Moothattu S, Balasubramanian K. Metal free, sunlight and white light based photocatalysis using carbon quantum dots from *Citrus grandis*: A green way to remove pollution. Solar Energy. 2018;**169**:120-127

[107] Zhu Z, Yang P, Li X, Luo M, Zhang W, Chen M, et al. Green preparation of palm powder-derived

carbon dots co-doped with sulfur/chlorine and their application in visible-light photocatalysis. Spectrochimica Acta Part A: Molecular and Biomolecular Spectroscopy. 2020;**227**:117659. DOI: 10.1016/j.saa.2019.117659

[108] Tang C, Cheng M, Lai C, Li L, Yang X, Du L, et al. Recent progress in the applications of non-metal modified graphitic carbon nitride in photocatalysis. Coordination Chemistry Reviews. 2023;**474**:214846

[109] Zhou J, Jia Q, Wang L, Zhao Y, Ma X, Gong L, et al. Highly efficient and selective photocatalytic CO2 reduction using MIL-125 (Ti) and based on LiFePO4 and CuO QDs surface–interface regulation. Catalysis Science & Technology. 2022;**12**:5152-5161. DOI: 10.1039/D2CY00917J

[110] Wang F, Hou T, Zhao X, Yao W, Fang R, Shen K, et al. Ordered macroporous carbonous frameworks implanted with CdS quantum dots for efficient photocatalytic CO2 reduction. Advanced Materials. 2021;**33**:2102690. DOI: 10.1002/adma.202102690

[111] Yan X, Jin Z. Interface engineering: NiAl-LDH in-situ derived NiP2 quantum dots and Cu3P nanoparticles ingeniously constructed pn heterojunction for photocatalytic hydrogen evolution. Chemical Engineering Journal. 2021;**420**:127682. DOI: 10.1016/j.cej.2020.127682

[112] Hu X, Yu Y, Chen D, Xu W, Fang J, Liu Z, et al. Anatase/rutile homojunction quantum dots anchored on g-C3N4 nanosheets for antibiotics degradation in seawater matrice via coupled adsorption-photocatalysis: Mechanism insight and toxicity evaluation. Chemical Engineering Journal. 2022;**432**:134375. DOI: 10.1016/j.cej.2021.134375

[113] Samadi-Maybodi A, Farzinia H. Application of CdS QDs incorporated in magnetized powder activated carbon for degradation of some dyes: Photodegradation process and comprehensive catalytic and spectroscopic studies. Journal of Photochemistry. 2018;**357**:103-117. DOI: 10.1016/j.jphotochem.2018.01.019

Chapter 2

Quantum Materials for Additive Manufacturing Applications

Ning Tu, Chengbin Kang, Mingjie Li and S.W. Ricky Lee

Abstract

Quantum materials' limited emission spectrum and easily adjustable color through particle size modification make them a viable option for the next generation of displays. The emission spectrum of quantum materials is sharp and pure, which makes quantum materials ideal for display applications. As display technology advanced, self-emitting display technology eventually replaced liquid crystal display (LCD). Researchers design different types of RGB pixels in the self-emitting display area to achieve the best possible visual impact. However, different types of pixels need quantum color conversion films with various patterns. Additive manufacturing offers a novel method for quicker prototyping of red, green, and blue (RGB) pixels with a faster iteration cycle. With the additive manufacturing technique, especially the inkjet printing method, the sample is not in contact with the surfaces; only the essential components are dispensed and deposited there. The additive manufacturing technique generally reduces sample damage or containment and material waste. This chapter introduces inkjet-printing quantum materials for high-resolution display applications.

Keywords: quantum dots, quantum perovskite, additive manufacturing, inkjet printing, display

1. Introduction

1.1 Emerging quantum dots for display applications

When the material's size is narrowed down to the nano size, the interaction at the atomic and subatomic scale makes the quantum materials the magic material. The solid material loses its conventional order and takes on wavelike properties on the atomic and subatomic scale. In this situation, quantum phenomena such as interface, tunneling, fluctuations, entanglement, and topological effects are found. All these quantum phenomena are indeed a fascinating and nearly spooky field; over the past decade, we have seen a significant change in the knowledge and utilization of multiple applications of quantum physics, especially when utilizing quantum materials' remarkable optical properties in display areas.

Among the quantum materials, quantum dots and perovskite nanoparticles have been widely studied over the years. At the same time, quantum dots (QDs) were first discovered in the 1980s by Louis E. Brus [1]. At that time, QDs only had a core without any passivation included. Following years of research, scientists discovered that covering the light-emitting core of QDs with a monolayer or several monolayers of a larger band gap material shell can significantly improve the optical properties of QDs [2–5]. QDs are frequently covered with organic ligands, commonly amine-based compounds such as ODA OR TOPO. The primary objective of this coating is twofold: to protect the shell surface from unfulfilled chemical interactions and to enable the dispersion of QDs in different solvents. **Figure 1(a)** is a typical structure of QDs, where the core of QDs is covered with shell and stabilizer (organic ligands).

Quantum confinement is one of the most unique properties of QDs. De Broglie's theory states that minuscule particles like electrons and holes can act like waves [6]. On the other hand, classical waves, like electromagnetic waves (light), can behave like particles (photons). This is called wave-particle duality. Thus, we can find that the difference of energy states is inverse to the square of the standing wave/diameter of QDs. Therefore, the different QD sizes will have different energy levels, which lead to different wavelengths. **Figure 1(b)** is a photograph of the Cd/Se QD solution in multiple sizes. When the size of QD is 2.5 nm, the emitting wavelength is 480 nm, while the diameter of QD is 6.3 nm, the emitting wavelength is 640 nm. It can be found that the smaller the size of QD, the shorter wavelength it emits.

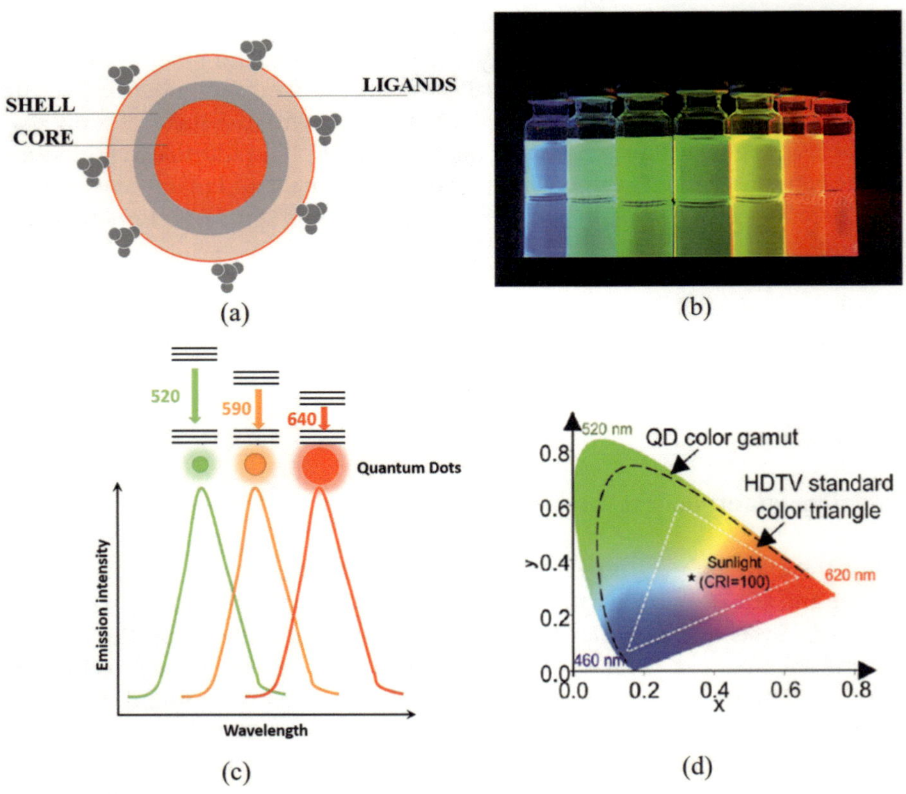

Figure 1.
(a) Illustration of QD structure. (b) The illustration of Cd/Se QDs solution of various sizes with the responsive emitting wavelength. (c) The emission spectrum of QDs with different sizes. (d) CIE chromaticity diagram of HDTV vs. QDs. Figure reproduced with permission from (c) Ref. [2] ©2017 Royal Society of Chemistry. Figure reproduced with permission from (d) Ref. [3] © 2018 Springer Nature.

Besides, the emitting wavelength of QDs can be changed by only changing the size of QDs. QDs are known to have a very narrow emission band. The full-width half-maximum (FWHM) value of the emission spectrum is applied to analyze the emission band; the smaller value of FWHM means a sharper emission spectrum peak, which results in a more saturated color [7]. **Figure 1(c)** is the emission spectrum of QDs with different sizes. The emission spectrum is narrow and sharp, with a smaller value of FWHM. **Figure 1(d)** illustrates that the color gamut value with QDs is much higher than HDTV's standard color gamut value. The white dashed triangle clearly shows a smaller spanning color gamut area than the black dashed line, which shows that QDs can reach color coordinate points with wavelengths ranging from 460 to 620 nm.

1.2 Emerging perovskite quantum dots for display application

Besides quantum dots, perovskite has attracted attention due to its good optoelectronics properties and the various types of these kinds of materials. Perovskite, a mineral composed of calcium titanium oxide ($CaTiO_3$), was identified. A perovskite is a substance that has a crystal structure that can be described by the formula ABX_3. Due to the unique property, based on the radiative recombination, various types of materials together with different shapes like 0D nanocrystals (NCs), 1D nanowires (NWs), and 2D nanoplates (NPLs) are emerging for unique scenarios. Various types of perovskites have been applied in different optoelectronic applications such as solar cells, photodetectors, light-emitting diodes (LEDs), high-power lasers, and biosensors. Here, we introduce two methods to prepare the perovskite QDs: the hot injection method and the ligand-assisted reprecipitation method.

The hot injection technique was the first developed for the synthesis of II-VI group colloidal cadmium selenide (CdSe) QDs [8], and then, it has been widely used for synthesizing different colloidal quantum dots nanoparticles [9–12]. During the hot injection synthesis, a cold stock solution with precursors is fast injected into the solution at a high temperature, which contains surfactant and a high boiling point solvent. The key to success is to maintain very little induction time between solution injection and particle precipitation. By this approach, bright colloidal perovskite with a defined size can be achieved in a supersaturated solution, which brings the development of large-scale application of QDs. In 2015, Protesescu et al. first proposed a highly efficient all-inorganic perovskite. These perovskites are highly luminescent colloidal quantum dots with narrow bandwidth and cube sizes around 4–15 nm. The materials composition with cesium lead halide perovskites ($CsPbX_3$, X = Cl, Br, and I or mixed halide systems Cl/Br and Br/I) has been produced by economic synthesis methods. By specific tuning of materials composition with quantum confinement effect. During the synthesis process, highly luminescent colloidal $CsPbX_3$ QDs were created by injecting a Cs-oleate precursor at temperatures ranging from 140 to 200°C. This injection induced the formation and enlargement of the QDs, which have adjustable emission spectra. The wavelength of the emitted light can be tuned across the entire visible range, making these QDs advantageous for display applications. The high luminescent $CsPbX_3$ QDs can achieve a photoluminescence quantum yield (PLQY) over ~90% for efficient display performance (**Figure 2**) [13].

Another synthesis method is the ligand-assisted reprecipitation method. High photoluminescence quantum yields (PLQYs) and intense electroluminescent (EL) emissions may only be attained with high excitation fluencies or current density due to nonradiative routes facilitated by sub-band defect states. The ligand-assisted

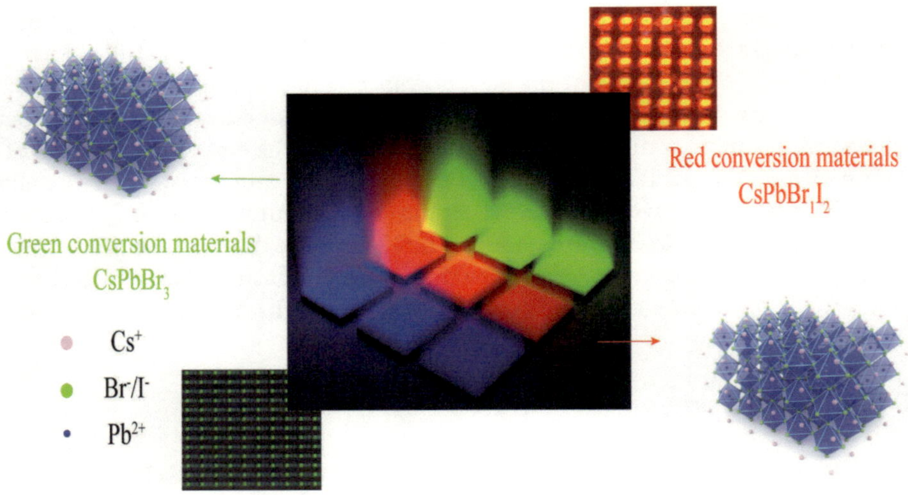

Figure 2.
Illustration of synthesized perovskite for display applications.

reprecipitation technique is a powerful approach to the synthesis of organometal halide perovskite nanocrystals [14]. In 2015, Zhang et al. developed an approach by ligand-assisted reprecipitation strategy (LARP) to synthesize efficient colloidal $CH_3NH_3PbX_3$ (X = Br, I, Cl) quantum dots with 70% quantum yield at room temperature and low excitation fluencies, which can be applied for display application. The LARP synthesis involves the combination of CH3NH3PbBr3 precursors in a solvent (N-dimethylformamide, DMF) and mixing it with a vigorously stirred solvent (toluene, hexane, etc.) that has long-chain organic ligands. This process results in colloidal nanoparticles' controlled formation through the precursors' crystallization. The intense increase of exciton binding energy in QDs and their proper surface passivation account for the emission enhancements in $CH_3NH_3PbX_3$QDs [15]. **Figure 3** is the synthesis method of the LARP strategy with the basic structure of the designed perovskite QDs.

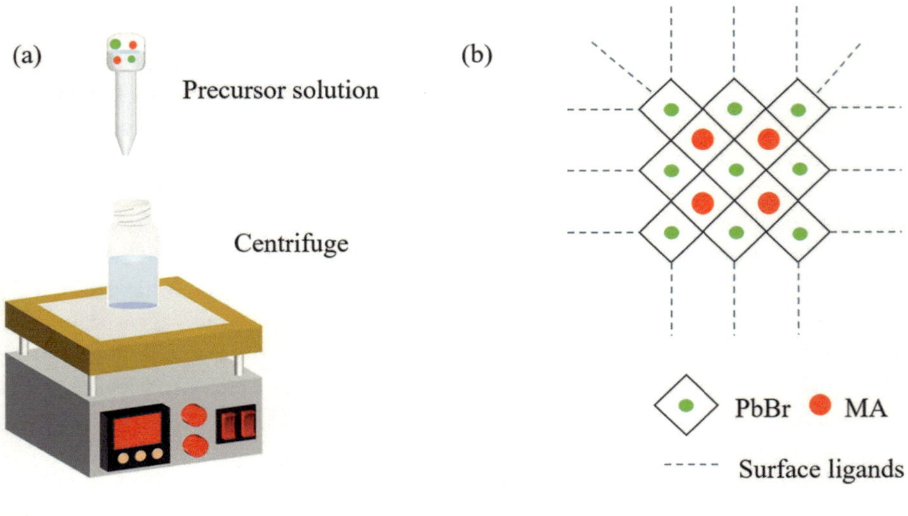

Figure 3.
(a) Synthesis method of LARP strategy. (b) The basic structure of the designed perovskite.

2. Inkjet printing quantum materials for display applications

Printing quantum materials for display applications has been widely studied in recent years due to its low manufacturing cost and fast prototyping. Traditional techniques such as photolithography, microcontact printing, and nanoimprinting have drawbacks such as high cost, severe causticity, unpredictable transfer printing, and poor machining efficiency. On the other hand, inkjet printing, which was previously used to create full-color displays, has a straightforward procedure, is inexpensive, highly automated, and allows for selective patterning without the need for a mask [16]. Here, we review excellent high-resolution patterning methods that have been developed and are compatible with existing device manufacturing processes to achieve good optical properties.

The history of printing technology dates back more than 1700 years—furthermore, the majority of printing techniques deposit material on the substrate by use of several physical events. Relief is used in flexography and gravure, surface energy is used in offset lithography, masking is used in screen printing, material is replaced by high force and pressure in embossing, material is removed by laser ablation, and material is released by excessive force and energy in laser transfer [17]. A non-contact, mask-free, additive method for depositing films with any design is inkjet printing [7, 18, 19].

2.1 Inkjet printing

The inkjet printing method is divided into continuous and drop-on-demand (DOD). Within the DOD group, there are two different types of inkjet technologies: piezoelectric inkjet technology and thermal inkjet technology, often known as bubble jet technology. Regarding the droplet manufacturing method, thermal inkjet technology differs from piezoelectric inkjet technology. All the methods' guiding concepts are shown in **Figure 4**.

The continuous inkjet printing method employs high-voltage deflection plates to redirect the steam of droplets electrically. Unsettled droplets are gathered by a gutter and then recirculated again, which is shown in **Figure 4(a)**. The piezoelectric inkjet

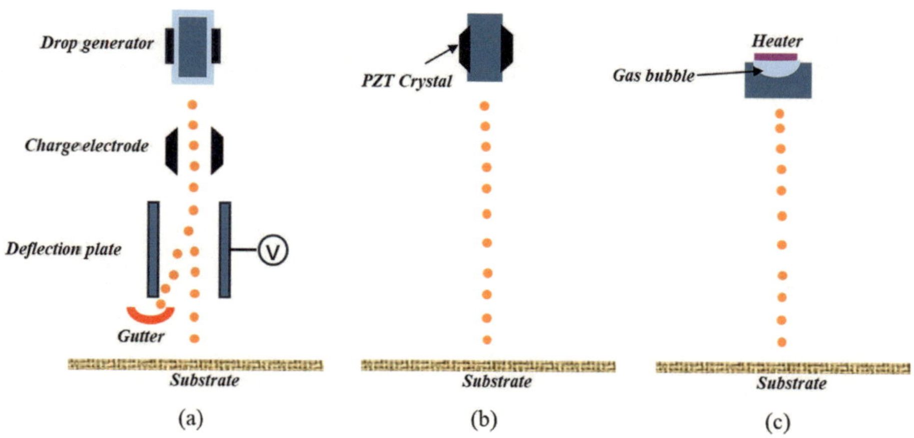

Figure 4.
Schematic of three main categories of inkjet printing technology. (a) Continuous inkjet printing (b) Piezoelectric inkjet printing (c) Thermal inkjet printing.

printing process utilizes a chamber containing a piezoelectric crystal that undergoes expansion or contraction in response to an applied bias. The deformation of this piezoelectric material generates a pressure wave that compels the material to expand and exit via an aperture, which refers to **Figure 4(b)**. The thermal inkjet printing technology utilizes the heating of a vapor bubble to generate an enlarged bubble, generating an acoustic wave and ultimately expelling a droplet from the chamber [20]. **Figure 4(c)** is the schematic of thermal inkjet printing. Thermal inkjet printing is widely used for graphics and low-end color printing. Meanwhile, the piezoelectric inkjet printer has a higher level of sensitivity toward the ink used. It can print temperature-sensitive biological samples or inks with very low viscosity. As a result, it is the favored choice for printing functional materials. Furthermore, piezoelectric print heads are more capable of generating droplets of varying sizes compared to thermal inkjet printers.

The DOD piezoelectric printing method is widely applied in mass production of OLED and QD panel fabrication [19]. However, current DOD piezoelectric printing still needs to improve on two fundamental issues when stepping toward high-resolution pixels. First, the ink composition and ejection will affect the printing quality [18, 21, 22]. The precise control over the small jetting nozzles to prevent the dislocation of droplets is the critical parameter to ensure delicate patterning, which is strongly related to the jetting parameters. The inverse Ohnesorge number defines the jetting parameters:

$$Z = \frac{\sqrt{\sigma \rho d}}{\gamma} \quad (1)$$

Here, the Ohnesorge number depends on the ink viscosity (γ, mPa s), surface tension (σ, N m^{-1}), density (ρ, g cm^{-3}), and printer nozzle diameter (d, μm); precise printing can be only realized by carefully designed jetting parameter with $1 < Z < 14$. With a Z number larger than 15, the satellite droplets will come out together with the primary drop. Optimized solvents with binary or ternary solution systems have been designed for optimal printing conditions [23, 24].

Another problem lies in the formation of the coffee ring effect during solvent evaporation due to the different evaporation rates in the droplet profile. **Figure 5** is the schematic of the coffee ring effect and the Marangoni flow. After the droplet lands on the substrate, the evaporation rate is greater near the edges of the droplet than near the center. Consequently, there is a greater dispersion of freedom from the central region to the periphery and from a capillary from the center to the edge. This capillary flow from the center to the edge brings the quantum materials to settle to the droplet edge and form the coffee ring effect. One of the methods to reduce the coffee ring

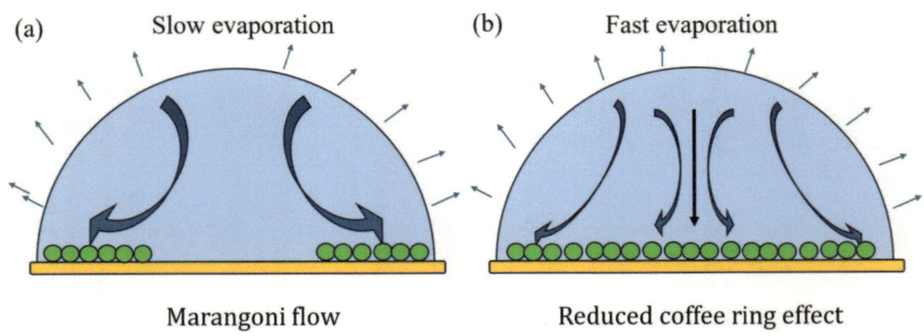

Figure 5.
Schematic of coffee ring effect and Marangoni flow.

effect is by adding a co-solvent into the quantum dots ink and creating a Marangoni flow from the edge to the center, against the capillary flow. In this way, the quantum dots distribute uniformly on the substrate without the coffee ring effect. Once the co-solvent has higher vapor pressure than the original solvent, and the surface tension of the co-solvent is also higher than the original solvent, and then the induced Marangoni flow would reduce the coffee ring effect. **Figure 6** is the 3D morphology image of QD dots on the substrate with different ratios of co-solvent. In **Figure 6**, toluene was applied as the co-solvent in the decane QDs solution. It can be seen that when the co-solvents adding ratio is to a proper value, the coffee ring effect is solved [25, 26]. This is one step closer to inkjet printing high-resolution QD display.

Besides directly printing QDs on the surface of the desired substrate, researchers also inkjet print the perovskite QDs into a polymer layer, known as in-situ inkjet printing (ISIP) strategy. SHI et al. utilized this technology to fabricate perovskite QDs patterns by printing perovskite precursor solution onto a polymeric layer, with PLQY around 80%. The high-resolution microdots array can be easily fabricated on the surface of the polymeric coating. This ISIP strategy paved the way for fabricating large area, bright luminescent, and color tunable micro-pattern $APbX_3$ (A = MA, FA, Cs, X = Cl, Br, I) PQDs, showing more advanced than the traditional inkjet printing [27], which requires carefully optimized ink composition and narrow process time window. **Figure 7** is the process of in-situ inkjet printing of perovskite QDs polymer film.

2.2 Aerosol jet printing

Besides the inkjet printing method, aerosol jetting is also widely used in quantum materials deposition applications. The aerosol jetting stands out due to its high printing resolution and delivery speed. **Figure 8** is the printing mechanism of aerosol jetting. As illustrated in **Figure 8**, the concentrated aerosol ink is generated in chamber ① using atomization techniques. Ultrasonic waves are used to create an aerosol from a modest amount of low-viscosity inks (1–500 cP) [28]. Next, the appropriate aerosol ink is conveyed to part ② using an inert carrier gas. The aerosol ink is then further transported to the focusing and deposition section of the aerosol jetting

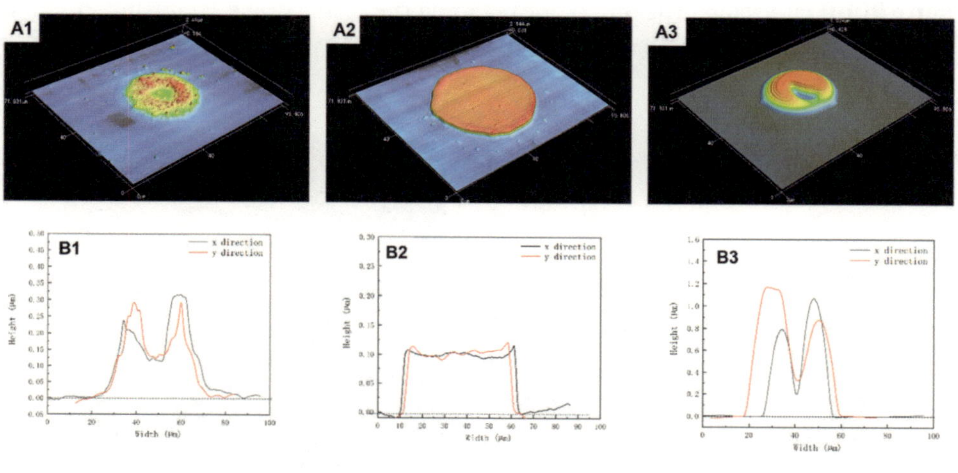

Figure 6.
The 3D morphology image of polymer QDs dots with different co-solvent ratios. (A) 3D morphology image of QDs dots. (B) QDs dots cross-section profile in x-direction and y-direction. Figure reproduced with permission from Ref. [25] ©2023, OAPA.

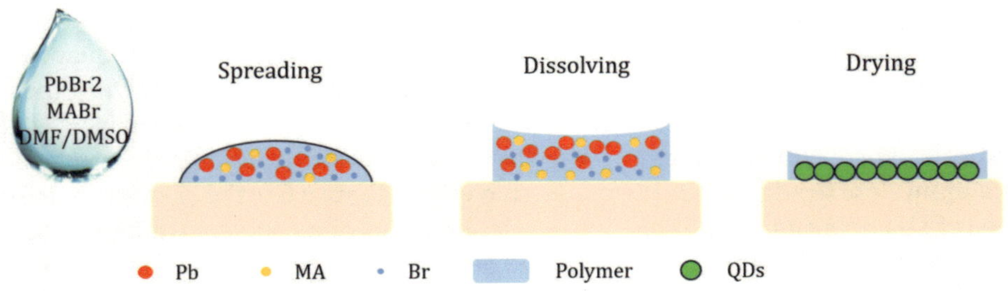

Figure 7.
Process of in-situ synthesis of perovskite polymer.

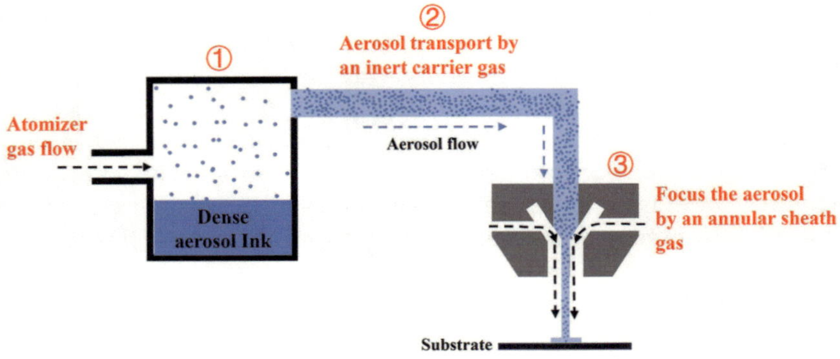

Figure 8.
Illustration of aerosol jetting printing mechanism.

part ③. During the focusing and deposition process, an annular sheath gas is utilized to concentrate the aerosol gas, and the aerosol ink is subsequently applied onto the desired substrate. Kuo et al. demonstrated aerosol jet printing high-resolution CdSe/ZnS QDs color conversion film for full-color μLED display applications [29]. The research applied a photoresist (PR) mold to resolve the coffee ring effect problem. **Figure 9(a-d)** illustrates the process of reducing the coffee ring effect using PR mold. The RGB pixels display independent emission is shown in **Figure 9(e)**.

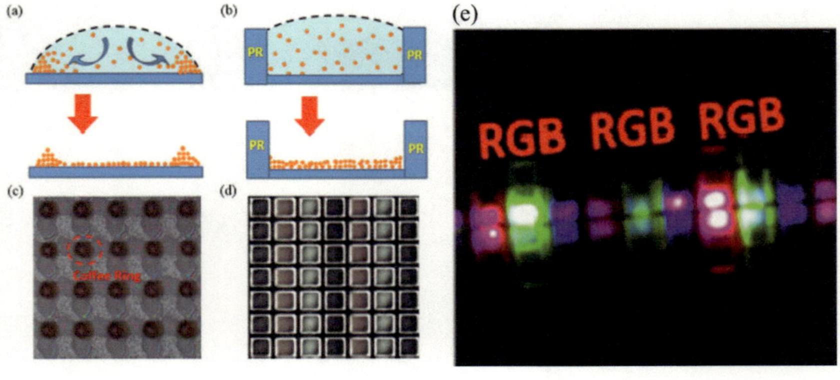

Figure 9.
(a)-(d) Process to prevent coffee ring effect by PR mold. (e) Illustration of RGB pixel emitting. Figure reproduced with permission from (a)-(e) Ref. [29] © 2017, Photonics Research.

3. Current and future trends in printing quantum materials in display applications

Quantum materials are functional semiconductor nanomaterials with special optical properties. Meanwhile, quantum materials are suitable for the solution process, which makes printing quantum materials a promising method in display manufacturing technology. In display technology, quantum materials have a sharp emission spectrum with tunable color by just changing the size. LCD display first applies quantum dots for color enhancement purposes, which further enhances the color performance and luminance value of LCD display. The market penetration of LCD displays equipped with QDs was estimated at about 5% in 2021.

Display markets are pursuing higher resolution, lower cost, and higher efficiency. Compared with traditional LCD TV panels, Samsung's QD-OLED was first launched in 2022. With the quantum materials applied in OLED display, mini-LED display, and micro-LED display technology, using printing methods to deposit the quantum materials becomes a trend in new display technology. The inkjet printing method is first introduced to deposit quantum materials. However, through inkjet printing technology, there are three dominant challenges to obtaining high-quality printing results, which include the satellite effect problem affecting the jetting process, the coffee ring effect problem affecting the drying process once quantum materials deposit on the substrate, and stable layer to layer printing process involving the stable printing process [30, 31]. For the aerosol jetting method, the aerosol steam during the printing process always comes with satellite dots. However, these satellite dots cannot be controlled or predicted, affecting the display's printing quality. Thus, people use masks to confine the printing area and control the printing process. Perovskite-based display systems, particularly those utilizing mini-LED/Micro-LEDs boosted by perovskite materials, will significantly improve display technology by offering superior quality with increased contrast, reduced power consumption, longer lifespan, and faster response times. Nevertheless, a perovskite-based panel is still needed to enhance the longevity [32, 33].

The remaining challenge of printing quantum materials for display applications is to achieve ultra-high-resolution patterning of quantum materials pixels. High-resolution displays require pixels that are more petite than the human eye's resolution limit, about 300 pixels per inch (PPI) for average viewing distance. However, conventional methods of quantum materials patterning, such as inkjet printing and aerosol jetting, are limited by the nozzle size, the surface tension of the ink, and the substrate roughness, resulting in pixel sizes that are larger than 10 micrometers or about 100 PPI. **Figure 10** shows the application of future trends in mini-LED/micro-LED with different panel sizes and resolutions. From digital display to AR, the panel size shows a downward trend. Meanwhile, the PPI shows a growing trend. Thus, to fulfill the requirement, a new printing method with a nanopipette was developed [34]. Different colors of quantum materials can be printed by varying the chemical composition of the ink. The nano-printing technique can produce quantum materials pixels with sizes as small as 100 nanometers, or about 2550 PPI, which is about 400 percent higher resolution than the latest high-end smartphones.

Another method of quantum materials patterning is based on using a nanocrystal mask, a thin film of quantum materials nanocrystals that can be selectively etched by a laser beam. Scanning the laser beam over the nanocrystal mask allows a pattern of quantum materials pixels to be transferred to the underlying substrate. The nanocrystal mask technique can produce quantum materials pixels with sizes as small as 200 nanometers, or about 1270 PPI, comparable to a retina display's resolution.

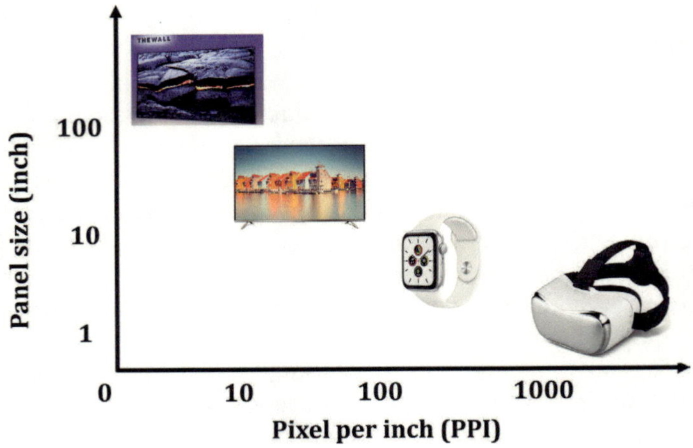

Figure 10.
Illustration of future trend mini-LED/micro-LED application with different panel sizes and resolutions.

4. Conclusion

In conclusion, high-resolution quantum materials printing is a promising technique for display applications, as it can offer superior performance and functionality over conventional materials and methods. Future research and development trends would focus more on improving the lifetime of non-toxic quantum materials, stable printing ink development, and nano-printing techniques with higher stability and resolution. The commercial viability and widespread adoption of printing quantum materials for display applications will be achieved with the realization of stable nano-printing techniques with stable quantum materials ink.

Acknowledgements

The Foshan government funded this research through a grant to HKUST Foshan Research Institute for Smart Manufacturing, the project of the Hetao Shenzhen-Hong Kong Science and Technology Innovation Cooperation Zone (HZQB-KCZYB-2020083), and the Innovation and Technology Commission in Hong Kong under Grant (ITS/010/21).

Conflict of interest

The authors declare no conflict of interest.

DOI: http://dx.doi.org/10.5772/intechopen.1005629

Author details

Ning Tu[1,2,3], Chengbin Kang[4,5,6], Mingjie Li[4,5,6] and S.W. Ricky Lee[1,7,8,9]*

1 Department of Mechanical and Aerospace Engineering, The Hong Kong University of Science and Technology, Hong Kong, China

2 Foshan Research Institute for Smart Manufacturing, The Hong Kong University of Science and Technology, Hong Kong, China

3 Electronic Packaging Laboratory, The Hong Kong University of Science and Technology, Hong Kong, China

4 Department of Applied Physics, The Hong Kong Polytechnic University, Kowloon, Hong Kong, China

5 Shenzhen Research Institute, The Hong Kong Polytechnic University, Shenzhen, Guangdong, China

6 Photonics Research Institute, The Hong Kong Polytechnic University, Kowloon, Hong Kong, China

7 Smart Manufacturing Trust, The Hong Kong University of Science and Technology (Guangzhou), Guangzhou, China

8 HKUST Shenzhen-Hong Kong Collaborative Innovation Research Institute, Shenzhen, China

9 HKUST LED-FPD Technology R and D Centre, Foshan, Guangdong, China

*Address all correspondence to: rickylee@ust.hk

References

[1] Bogue R. Quantum dots: A bright future for photonic nanosensors. Sensor Review. 2010;**30**:279-284. DOI: 10.1108/02602281011072143

[2] Ren DH, Wang B, Hu C, You Z. Quantum dot probes for cellular analysis. Analytical Methods. 2017;**9**:2621-2633. DOI: 10.1039/c7ay00018a

[3] Choi MK, Yang J, Hyeon T, Kim D-H. Flexible quantum dot light-emitting diodes for next-generation displays. NPJ Flexible Electronics. 2018;**2**:1-14. DOI: 10.1038/s41528-018-0023-3

[4] Wilson WL, Szajowski PF, Brus LE. Quantum confinement in size-selected, surface-oxidized silicon nanocrystals. Science. 1993;**262**:1242-1244. DOI: 10.1126/science.262.5137.1242

[5] Yang J, Choi MK, Kim DH, Hyeon T. Designed assembly and integration of colloidal nanocrystals for device applications. Advanced Materials. 2016;**28**:1176-1207. DOI: 10.1002/adma.201502851

[6] Morgon NH. The behavior of the electron: An analysis of the Compton effect and the de broglie's relation. Quimica Nova. 2008;**31**:1869-1874. DOI: 10.1590/s0100-40422008000700046

[7] Dai XL, Deng YZ, Peng XG, Jin YZ. Quantum-dot light-emitting diodes for large-area displays: Towards the dawn of commercialization. Advanced Materials. 2017;**29**:1607022. DOI: 10.1002/adma.201607022

[8] Murray CB, Norris DJ, Bawendi MG. Synthesis and characterization of nearly monodisperse CdE (E = sulfur, selenium, tellurium) semiconductor nanocrystallites. Journal of the American Chemical Society. 1993;**115**:8706-8715. DOI: 10.1021/ja00072a025

[9] Kang C et al. Quantum-rod on-chip LEDs for display backlights with efficacy of 149 lm W−1: A step toward 200 lm W−1. Advanced Materials. 2021;**33**:2104685. DOI: 10.1002/adma.202104685

[10] Prodanov MF et al. Thermally stable quantum rods, covering full visible range for display and lighting application. Small. 2021;**17**:2004487. DOI: 10.1002/smll.202004487

[11] Liao Z et al. Ultralow roll-off quantum dot light-emitting diodes using engineered carrier injection layer. Advanced Materials. 2023;**35**:2303950. DOI: 10.1002/adma.202303950

[12] Kang CB et al. Robust, narrow-band nanorods LEDs with luminous efficacy > 200 lm/W: Next-generation of efficient solid-state lighting. Small. Mar 2024;**27**:2311671. DOI: 10.1002/smll.202311671

[13] Protesescu L et al. Nanocrystals of cesium lead halide perovskites (CsPbX3, X = Cl, Br, and I): Novel optoelectronic materials showing bright emission with wide color gamut. Nano Letters. 2015;**15**:3692-3696. DOI: 10.1021/nl5048779

[14] Tang Y et al. Precursor solution volume-dependent ligand-assisted synthesis of CH3NH3PbBr3 perovskite nanocrystals. Journal of Alloys and Compounds. 2019;**773**:227-233. DOI: 10.1016/j.jallcom.2018.09.054

[15] Zhang F et al. Brightly luminescent and color-tunable colloidal

CH3NH3PbX3 (X = Br, I, Cl) quantum dots: Potential alternatives for display technology. ACS Nano. 2015;**9**:4533-4542. DOI: 10.1021/acsnano.5b01154

[16] Tseng H-Y. Scaling of inkjet-printed transistors using novel printing techniques. [thesis]. USA: UC Berkeley; 2011

[17] Casiraghi C, Jurchescu OD, Magdassi S, Su WM. Introduction to nanomaterials for printed electronics. Nanoscale. 2023;**15**:10480-10483. DOI: 10.1039/d3nr90072j

[18] Gao Y, Kang C, Prodanov MF, Vashchenko VV, Srivastava AK. Inkjet-printed, flexible full-color photoluminescence-type color filters for displays. Advanced Engineering Materials. 2022;**24**:2101553. DOI: 10.1002/adem.202101553

[19] Kant C, Mahmood S, Katiyar M. Large-area inkjet-printed OLEDs patterns and tiles using small molecule phosphorescent dopant. Advanced Materials Technologies. 2023;**8**:2201514. DOI: 10.1002/admt.202201514

[20] Korvink JG, Smith PJ, Shin D-Y. Inkjet-Based Micromanufacturing. Germany: Wiley; 2012. DOI: 10.1002/9783527647101

[21] Kang C, Zhou Z, Halpert JE, Srivastava AK. Inkjet printed patterned bank structure with encapsulated perovskite colour filters for modern display. Nanoscale. 2022;**14**:8060-8068. DOI: 10.1039/d2nr00849a

[22] Bastola A et al. Formulation of functional materials for inkjet printing: A pathway towards fully 3D printed electronics. Materials Today Electronics. 2023;**6**:100058. DOI: 10.1016/j.mtelec.2023.100058

[23] Chen M et al. High performance inkjet-printed QLEDs with 18.3% EQE: Improving interfacial contact by novel halogen-free binary solvent system. Nano Research. 2021;**14**:4125-4131. DOI: 10.1007/s12274-021-3352-9

[24] Tu N, Lo JCC, Lee SWR. Improved inkjet printing of polydimethylsiloxane droplets. IEEE. International Conference on Electronics Packaging (ICEP); 11-14 May 2022; Sapporo, Japan. IEEE; 2022. pp. 241-242

[25] Tu N, Lo JCC, Lee SWR. Development of uniform polydimethylsiloxane arrays through inkjet printing. Polymers. 2023;**15**:462. DOI: 10.3390/polym15020462

[26] Wei CT et al. A universal ternary-solvent-ink strategy toward efficient inkjet-printed perovskite quantum dot light-emitting diodes. Advanced Materials. 2022;**34**:2107798. DOI: 10.1002/adma.202107798

[27] Shi L et al. In situ inkjet printing strategy for fabricating perovskite quantum dot patterns. Advanced Functional Materials. 2019;**29**:1903648. DOI: 10.1002/adfm.201903648

[28] Tu N, Lo JCC, Lee SWR. Printing unifrom QDs polymer thin film for QLED applications. In: 70th IEEE Electronic Components and Technology Conference (ECTC). 3-30 June 2020; Orlando, FL, USA; 2020. pp. 1992-1998

[29] Lin HY et al. Optical cross-talk reduction in a quantum-dot-based full-color micro-light-emitting-diode display by a lithographicfabrication photoresist mold. Photonics Research. 2017;**5**:411-416. DOI: 10.1364/PRJ.5.000411

[30] Dayneko S et al. Application of CdSe/ZnS/CdS/ZnS core—Multishell quantum dots to modern OLED technology. Materials Today: Proceedings. 2016;**3**:211-215. DOI: 10.1016/j.matpr.2016.01.059

[31] Park YG et al. Defect engineering for high performance and extremely reliable a-IGZO thin-film transistor in QD-OLED. Advanced Electronic Materials. 2022;**8**:2270036. DOI: 10.1002/aelm.202101273

[32] Hu Z et al. Inkjet printed uniform quantum dots as color conversion layers for full-color OLED displays. Nanoscale. 2020;**12**:2103-2110. DOI: 10.1039/c9nr09086j

[33] Wu XG, Ji HL, Yan XL, Zhong HZ. Industry outlook of perovskite quantum dots for display applications. Nature Nanotechnology. 2022;**17**:813-816. DOI: 10.1038/s41565-022-01163-8

[34] Hengsteler J, Lau GP, Zambelli T, Momotenko D. Electrochemical 3D micro-and nanoprinting: Current state and future perspective. Electrochemical Science Advances. 2022;**2**:e2100123. DOI: 10.1002/elsa.202100123

Chapter 3

Quantum Complexity and Emulation of Black Holes

Lawrence B. Crowell

Abstract

In this paper the fundamental quantum states of the world are argued to be determined by the Riemann ζ function as zeros of the supersymmetric Hamiltonians. This is a sort of quantum-gravitation dressed state that is quasi-local. This is then a model for any system with quantum states, such as dressed quantum states found in quantum optical physics. The analogues from conformal gravitation, such as with $SL(4,\mathbb{F})$, for $F = \mathbb{R}$ or $\mathbb{C}$ carry to $SL(3,F)$ with QCD content and with $SL(2,\mathbb{F})$ corresponding to spin $\frac{1}{2}$ system and basic general relativity for $\mathbb{R}$ or $\mathbb{C}$ respectively. This then provides a general framework for quantum computing. In the case of a quantum dot the system is the simplest with $SL(2,\mathbb{R}) \rightarrow SU(2)$ as a Wick rotation. In this setting there is a general occurrence of dressed states, which nature configures that that may be engineered. Quantum dots as black hole models, regions that hold a quantum state as an instanton, are worked as a toy model. It is then possible to emulate black holes in the laboratory as a quantum computing element.

Keywords: conformal fields, quantum computing, Riemann ζ function, black holes, quantum optical dressed states, emulated black holes, quantum dots

1. Introduction

There has been a long history of considering natural systems according to devices, usually those recently developed. In the Renaissance the pump was considered to be most important with the motion of the aether in space. The clock and later steam power were later paradigms for how nature worked, such as Newton's clockwork universe and the thermodynamic world seen in the Carnot cycle and steam engine. The computer has come to fill a similar role today. Seth Lloyd has proposed that the entire universe is in fact a great quantum computer [1]. If the boundary and initial conditions of the universe can be established, then physical dynamics can be considered as a sort of computation. We can model physical systems on computers, even use classical computer to model quantum systems, and this leads to the conjecture that physics and even the entire universe is a type of computer.

A quantum dot is a sort of unit of a quantum electronic device, which could be a quantum computer. In what follows this will lead to a sort of regularization of the amount of quantum information such a system can hold. An elementary idea may be seen in the Rydberg states of the hydrogen atom. A naïve idea might be that it could

IntechOpen

hold an infinite amount of information, or information in some incredibly complicate superposition of electronic states. However, for high Rydberg numbers the Cauchy-like sequence of radial atomic states will make any distinguishability of them impossible. Thus for any real system with physical dressed states of this sort will have a finite information content. The quantum dot is then such a system. Towards the end this will be argued for conformal gravitation, as the Ryu-Takaynagi formula, with QCD and ultimately with a simple atomic state system or a quantum dot that holds electronic states.

Quantum Chromodynamics (QCD) of quarks and gluons is a decent model. Gluons have two QCD charges called colors and quarks carry one color. A gluon with a color red and anti-green that is absorbed by a quark that carries a green charge which swaps that charge for red. Similarly a quark with the color blue can emit a gluon with colors blue and anti-red, leaving the quark with the color red. The Gel-Mann matrices act as transformations with these colors and in dynamical equations, the coupled Dirac and Yang-Mills equations, this process can be modeled on a computer. Lattice gauge theory is a pseudo-temperature version of this sort of computation. We may think of QCD as a set of algorithmic rules for a quantum computer.

Care needs to be taken here. In particular with using classical computers to model quantum systems. This has various difficulties, where modeling the exponential $e^{-iHt/\hbar}$ is such that tiny overflow errors can explode. Digital filtering and various methods must be employed to prevent or limit this. This is in many ways a digital version of the Bell inequalities, where any sub-quantum wiring of a quantum system will lead to an exception to the violations of Bell inequalities. Further, this obviously cannot be a polynomial computation due to the exponential nature of the phase. This touches on the issues of superdeterminism of 't Hooft [2] and Hossenfelder and Palmer [3]. The invariant set theory inherent in this interpretation defines a Cantor set, or "monster," which is not computable. The metric geometry is p-adic [4], and there does not exist a global system for solution of these equation by the Matiyasevich theorem [5]. Issues of this nature, with computability and its role in physics, are discussed below. For this reason, the computing of quantum systems is best done with a quantum computer, which will doubtless be a frontier in the future.

Physics is necessary for laying down the stage or hardware of a computer. Solid State physics and quantum electronics are necessary for developing computing systems. The future will doubtless illustrate new architectures for computing, such as graphene electronics, spintronics, annealing and Boltzmann machines and of course quantum computers. As we push into these new frontiers the computing dynamics becomes closer to the actual types of systems used. This is certainly the case with quantum computers. In this we discuss the nature of quantum complexity with the connection to quantum black holes. Black holes are in some ways the ultimate quantum computer [6]. Herein these computing architectures are examined in the light of physics.

2. Quantum complexity and black holes

The Riemann ζ-function has been thought to reflect quantum mechanical spectra since this was suggested by Berry and Keating [7]. However, so far there has been no definitive demonstration of this. The ζ-function is by the famous ζ-function conjecture of Riemann determines distribution of prime numbers [8]. As yet this is an unproven theorem. The distribution of zero's for the ζ function is approximately equal to the logarithmic integral function $li(x) = \int_2^x \frac{dx'}{ln\ x'}$, where the function has been offset from $x' = 1$ where it diverges. The Riemann ζ-function occur on the line $z = \frac{1}{2} + iy$, where the distribution of y is approximated by this function. In this essay the role of the Riemann ζ-function and quantum complexity is examined.

The distribution of primes is approximately determined by a line with constant slope for $x \gg 2$. The quantum complexity of black holes obeys the logistic equation $\frac{df}{dx} = \frac{(K-f)f}{K}$ [9]. The solution to this differential equation is $f(x) = \frac{Ke^{x-x_0}}{1+e^{x-x_0}}$. In quantum complexity $z_0 = ln\ K$, which is the scrambling time. The volume of a region interior to the AdS-black hole spacetime evolves according to

$$\frac{dV}{dt} = \ell_{ads} AT \tag{1}$$

for time the Rindler or anchoring AdS time, and $\ell_{ads} = \frac{1}{\sqrt{R}}$ the AdS scale. The temperature T is a temperature connected to the black hole mass $T = \frac{1}{8\pi m}$ and is a Euclideanized time $\tau = \hbar kT$. This volume increases linearly with this time variable. The solution to the logistic function is approximately linear. At the start for $x - x_0 \simeq 0$ the function is approximately $f(x) \simeq \frac{1}{2}e^{(x-x_0)/2}$. Further, for $x \gg x_0\, f(x) \simeq 1$. The sigmoid curve in between is approximately linear. This is the final state where quantum complexity is maximal. The expansion of the natural logarithm around a value $a > 1$ is to linear approximation $ln\ (x) \simeq 1 + \frac{x-a}{a}$. The ofset logarithmic integral function $\int_a^x \frac{dx'}{x'} \simeq \frac{a}{x}$. This obeys the logistic differential equation for an expansion of the logarithm around $a \gg 1$ so the offset logarithmic integral is $\int_a^x \frac{dx'}{ln\ x'} \simeq x - a$. This obeys the logistic equation integration constant $c = K$. This assumes a deviation small from the offset constant a, but since this is arbitrary it is evident the function is approximately constant and obeys the logistic equation near the inflection point of the sigmoid curve.

This is then a physical argument for why the ζ-function plays a role in the complexity of black hole quantum states. Entropy S determines complexity $C = e^S$, where entropy and action have correlated meanings, and the complexity can be measured by logistic functions that approximate the ζ-function.

3. Quantum complexity and the Schwarzian

Kalosh et al. studied the conformal black hole [10] with the approximate form of the Schwarzschild metric near $r = 0$. The metric is

$$ds^2 = F(r)dt^2 - F(r)^{-1}dr^2 - r^2\Omega^2, \tag{2}$$

For $F(r) \simeq \frac{2m}{r}$. Rescaling the metric with $ds^2 \to F(r)^3 ds^2$ gives the effective Lagrangian

$$\mathcal{L} = \sqrt{\left(\frac{2m}{r}\right)^4 - \left(\frac{2m}{r}\right)^2 u_r^2 - \left(\frac{(2m)^3}{r}\right) L^2} \tag{3}$$

with $L = \frac{d\Omega}{dt}$ the angular momentum. A small change in variables with $m \to \mu m$, μ a test mass, so $u_r \to p_r$ gives

$$\mathcal{L} = \left(\frac{2m}{r}\right)^2 \sqrt{(r^2p_r^2 - 2mr^3L^2)/4m^2 - \mu^2}. \tag{4}$$

From this there is the Hamiltonian

$$H = \left(\frac{2m}{r}\right)^2 \sqrt{(r^2p_r^2 + 2mr^3L^2)/4m^2 + \mu^2}. \tag{5}$$

This Hamiltonian may be transformed into

$$H = \frac{p_r^2}{\alpha} + \frac{\beta}{\alpha r^2}, \tag{6}$$

for $\alpha = \sqrt{(\mu^2 + r^2p_r^2 + 2r^3L^2)/4m^2}$ and $\beta = 2m^2\mu^2 + mr^3L^2$. This connects with quantum field theory for $L^2 = \ell(\ell+1)$. In the limit that $m \to \infty$, but $m\mu$ remaining constant this defines a non-relativistic Hamiltonian for the conformal group

$$H = \frac{p_r^2}{2\mu} + \frac{\beta}{r^2}. \tag{7}$$

The equations of motion are Newtonian, and we may consider this the case where the black hole is so huge that geodesic separation or "gravity" is very small.

The metric $\left(\frac{2m}{r}\right)^2 ds^2$ gives the Lagrangian above which may be written with

$$\mathcal{L} = \sqrt{\frac{(f')^4 - (f')^2u_r^2 - \frac{(2m)^2}{r}L^2}{\alpha}}. \tag{8}$$

for $\alpha = \sqrt{(\mu^2 + r^2p_r^2 + 2r^3L^2)/4m^2}$ and $\beta = 2m^2\mu^2 + mr^3L^2$. This connects with quantum field theory for $L^2 = \ell(\ell+1)$. In the limit that $m \to \infty$, but $m\mu$ remaining constant this defines a non-relativistic Hamiltonian for the conformal group

$$H = \frac{p_r^2}{2\mu} + \frac{\beta}{r^2}. \tag{9}$$

The equations of motion are Newtonian, and we may consider this the case where the black hole is so huge that geodesic separation or "gravity" is very small.

The Virasoro algebra is defined with a basis $L_m = e^{im\theta}\partial_\theta$, where it is not hard to see the commutator $[L^m, L^n] = i(n-m)L^{m+n}$ holds. This is the Witt algebra. There are an $m+n$ possible number of commuting vectors which give the same result. To eliminate this degeneracy a kernel is used which results in the Virasoro algebra

$$[L^m, L^n] = (m-n)L^{m+n} + (c/12)(m^3 - m)\delta_{m+n,0}. \tag{10}$$

This is the fundamental algebra of the bosonic string states. The state operators (L_{-1}, L_0, L_1) form a closed algebra $SL(2,\mathbb{R}) \simeq SU(1,1)$. This is a system of boosts in special relativity. This means the system has an invariant given by a function $\psi(x)$ such that the Schwarzian

$$S(\psi) = \psi_{xxx}/\psi_x - (3/2)(\psi_{xx}/\psi_x)^2 \tag{11}$$

is invariant $S(\psi') = S(\psi)$ under $SL(2,\mathbb{R})$ transformations

$$\psi' = \frac{a\psi + b}{c\psi + d} \tag{12}$$

This function $\psi(x)$ corresponds to the states (L_{-1}, L_0, L_1) with the $SL(2,\mathbb{R})$ transformations of this function

$$\psi' = a\psi + b, \psi' = a\psi + b\psi, \psi' = \psi + c\psi^2 \tag{13}$$

define the elementary generators

$$P = i\partial_\psi.D = i\psi\partial_\psi, K = i\psi^2\partial_\psi, \tag{14}$$

For translation, dilatation, and special conformal transformation acting upon the form of the field $\psi(x)$. This closed algebra means the orbit space is bounded and there is no runaway fountain of radiation that results.

The Schwarzian derivative is equivalent to an invariant Maurer–Cartan one–forms, which make it $SL(2,\mathbb{R})$ invariant. This invariance means a Lagrangian form is possible [11]. The general group element is $g = e^{iHt}e^{i\psi(t)P}\ e^{i\sigma(t)K}e^{i\rho(t)D}$ where it is clear that $g^{-1}dg = i(Hdt + Pd\psi + Kd\sigma + Dd\rho)$ or $g^{-1}dg = \left(Hdt + P\omega_\psi + K\omega_\sigma + D\omega_r ho\right)$. With $t' = t + \delta t$ the one-forms are

$$dt, \omega_P = \dot{\psi}e^{-\rho}dt, \omega_k = e^\rho\left(\dot{\sigma} + \sigma^2\dot{\psi}\right)dt, \omega_D = (\dot{\rho} - 2\sigma\dot{\psi})dt \tag{15}$$

The imposition of constrants can reduce the number of independent variables. We impse $\omega_P - a\omega_H = 0$ and $\omega_D + 2b\omega_H = 0$ the constant are such that $a = \dot{\psi}e^{-u}$ and $b = s\dot{\psi} - \ddot{\psi}/2\dot{\psi}$, or

$$s = \frac{b}{\dot{\psi}} + \frac{\ddot{\psi}}{2\dot{\psi}^2}, \tag{16}$$

and we then have a Lagrangian $\mathcal{L} = \dot{\psi}(\dot{s} + s^2\dot{\psi} - 2bs)$. In the Maurer-Cartan one-forms then have the coefficients

$$H = \dot{\psi}\left(\dot{s} + s^2\psi\right), P = \dot{s} + 2s^2\dot{\psi} - 2bs \tag{17}$$

$$D = \psi P - s\dot{\psi}, K = \psi^2 P + \dot{\psi}(1 - 2s\psi) + 2b\psi. \tag{18}$$

The Hamiltonian may be written then as

$$H = \frac{1}{2}\left(\frac{\dddot{\psi}}{\dot{\psi}} - \frac{3}{2}\frac{\ddot{\psi}^2}{\dot{\psi}^2}\right) + b^2, \tag{19}$$

which is the Schwarzian derivative. This is invariant under the linear fractional transformation $z \to \frac{az+b}{cz+d}$, which means this Hamiltonian is invariant under hyperbolic dynamics [12].

The "Schwarzian" interpreted as generating the extremal volume of the dual black hole interior, where the Hamiltonian of the conformal dynamics is determined by the upper bound on complexity, $C = e^S$ and the entropy $\simeq$ action defines the Hamiltonian $S = p\dot{q} - H$. This extremal limit is equal to the entropy measure in a hyperbolic wedge, which any particular wedge is set by the mass of the black hole. We may then think of this parameter as a type of gauge condition. The quadratic for Newman charged black holes is $\sqrt{m^2 - q^2}$ so that in general the Maurer–Cartan one–form pdx has a gauged qA as a Lagrange multiplier or constraint. This additional gauge condition is then associated with $m \to m + q/2$, for $q \ll m$. These dynamics are the same as complexity dynamics of a $1-d$ particle in a potential given by the angular momentum barrier. Eq. (8) has just such a barrier. The angular momentum in Eq. (8) maybe quantized as $\ell(\ell+1)\hbar$, and this is then a quantized barrier.

4. Riemann ζ-function realization

The Lagrangian for the mass-term $\mathcal{L} = \sqrt{g}m + qA$ holds for $E^M = (E^\mu, i, E^m)$ such that $E^{\mu,i}_m$ is a tetrad one-form. The metric is $g = E^a \otimes E^b \eta_{ab}$. The index μ is the $SL(2,\mathbb{C})$ indices $i = 1,2,3$ for $(1,2)$ the $SU(2)_R$ R-symmetry and 3 the $U(1)$ so the group is $U(2)$. The diffeomorphisms of the space is the group $SU(1,1) \times SU(2)$. The isometry group of $AdS_2 = SO(1,2)/SO(1,1)$ is $SO(1,2) = SU(1,1)$ and for $AdS_2 \times \mathbb{S}^2$ isometries are $SU(1,1) \times SU(2)$ This embeds into the super-isometry group $SU(1,1|2)$ for $N = 2$ supersymmetry and supergravity. The supergroup has four complex charges (8 real ones) in this $N = 2$ supersymmetry [10]. The conformal quantum numbers or charges (H,K,D) are determined by the supercharges

$$\left\{Q^\mu_i, \overline{Q}^j_\nu\right\} = -\frac{1}{4}\left(\delta^i_j\left(\tau^{ab}\right)^\mu_\nu L_{ab} + \delta^\mu_\nu\left(\tau^{a'b'}\right)^j_i L_{a'b'}. \quad (20)$$

Here $\tau_1 = \sigma_1$, $\tau_2 = \sigma_2$ and $\tau_3 = i\sigma_3$ for $SO(2,1)$ and the $\tau_i = \sigma_i$ for $SO(3)$. The conformal charges are then $H = -2(L_{02} + L_{01})$, $K = 2(L_{02} - L_{01})$ and $D = 2L_{21}$. The matrix $L_{\mu\nu} = \tau^{ab}_{\mu\nu}L_{ab}$ is comprised of the conformal generators

$$L_{\mu\nu} = \begin{pmatrix} iK & D \\ D & iH \end{pmatrix} \quad (21)$$

Such that for

$$\mathcal{Q} = \begin{pmatrix} S \\ Q \end{pmatrix}, \{\mathcal{Q}_\mu, \mathcal{Q}_\nu\} = L_{\mu\nu}. \quad (22)$$

The elementary supersymmetry operator may be thought of as removing a fermion and replacing it with a boson. This operator would be $q = a^\dagger b$ and its conjugate is $q^\dagger = b^\dagger a$ that removes a boson and replaces it with a fermion. The anticommutator

$\{q, q^\dagger\} = a^\dagger a + b^\dagger b$ is then the number operator or the number of particles. This is identified as the Hamiltonian. We may then consider the operators

$$Q = \begin{pmatrix} 0 & 0 \\ q & 0 \end{pmatrix}, Q^\dagger = \begin{pmatrix} 0 & q^\dagger \\ 0 & 0 \end{pmatrix}, \tag{23}$$

$$H = \begin{pmatrix} q^\dagger q & 0 \\ 0 & qq^\dagger \end{pmatrix} = \begin{pmatrix} H_- & 0 \\ 0 & H_+ \end{pmatrix}, \tag{24}$$

such that the number operator is $N = Tr(H)$. The vacuum state in this toy model is $|0_f, 0_b\rangle$ and clearly $q|0_f, N_b\rangle = 0$ and $q^\dagger|N_f, 0_b\rangle = 0$. This means $H_\pm$ annul the fermion and boson vacuum separately so $H_-|0_f, N_b\rangle = 0$ and $H_+|N_f, 0_b\rangle = 0$, and $H_-|1_f, N_b\rangle = N_b$ and $H_+|N_f, 1_b\rangle = N_f$.

The dilatation operator $D = x\frac{d}{dx}$ is an eigen-operator of a monomial x^s is such that $Dx^s = x\frac{dx^s}{dx} = sx^s$ [13]. This means a function of D gives the function of the eigenvalue so $Df(x^s) = f(s)x^s$. The dilatation operator can be the generator (derivation) of an operator such as $\sum_n (-1)^{n+1} ne^D$ or

$$\mathcal{O} = \sum_{n=1}^{\infty} (-1)^{n+1} e^{ln\,(x)D} \tag{25}$$

where the inverse operator must then be

$$\mathcal{O}^{-1} = \sum_{n=1}^{\infty} \frac{(-1)^{n+1}}{n} e^{ln\,(x)D} \tag{26}$$

It is not hard to see this operator acts on x^{-s} to give $\mathcal{O}x^{-s} = (1 - 2^{1-s}\zeta(s)x^{-s}$ and its inverse $\mathcal{O}^{-1}x^{-s} = (1 - 2^s\zeta(1-s)x^{-s}$. Further, the lowering and raising operators can be shown to be

$$a_\omega = |x|^{-i\omega/2}\mathcal{O}|x|^{-i\omega/2}, a_\omega^\dagger = |x|^{i\omega/2}\mathcal{O}^{-1}|x|^{i\omega/2}, \tag{27}$$

where upon the supersymmetric Hamiltonians $H_- = a_\omega^\dagger a_\omega$ and $H_+ = a_\omega a_\omega^\dagger$ or

$$H_- = |x|^{i\omega/2}\mathcal{O}^\dagger\mathcal{O}|x|^{-i\omega/2} \tag{28}$$

$$H_+ = |x|^{-i\omega/2}\mathcal{O}\mathcal{O}^\dagger|x|^{i\omega/2}. \tag{29}$$

The commutator of these is zero, which is evident from the fact they are constructed from the same dilatation operator.

These Hamiltonians act on quantum states $\psi \sim e^{i\Omega} = |x|^z$ in polar form with $z = a + ib$, where the lowering and raising operators act on these as

$$\begin{aligned} a_\omega|x|^z &= \left(1 - 2^{1-a+i(b-\omega/2)}\right)\zeta(a - i(b-\omega/2))|z|^{a+i(b-\omega/2} \\ a_\omega^{-1}|x|^z &= \left(1 - 2^{a+i(b+\omega/2)}\right)\zeta(1 - a + i(b+\omega/2))|z|^{a-i(b+\omega/2}. \end{aligned} \tag{30}$$

The Hamiltonians are then determined by the product of two ζ-functions [7]. The condition for the supersymmetric vacuum $H_\pm|0\rangle = 0$ then requires that $a = \frac{1}{2}$ and

$b+\omega/2\in\mathbb{Q}_p$ or $b+\omega/2\in\mathbb{Q}_p$. Here $\mathbb{Q}_p$ is the field of prime numbers. Then according to the Riemann conjecture on the zeros of the ζ -function one of these is zero and the Hamiltonian annuls the vacuum state.

The Hamiltonians may be computed

$$H_\pm - = \ |1-2^{1/2-i(b\pm\omega/2)}|\zeta(1/2-i(b\pm\omega/2))|^2|x|^{-i\omega}, \tag{31}$$

using $\zeta(1/2-x)^* = \zeta(1/2+ix)$. The eigenvectors are then

$$\psi_n^\pm = A_n|x|^{1/2+i\left(\left(n+\frac{1}{2}\right)\omega+(b\pm\omega/2)\right)}, \tag{32}$$

with the result, so the integration of with $z=x^{i\omega}$ (gives)

$$\int dx\psi_m^{\pm *}\psi_n^\pm = -\frac{i}{\omega}\oint dz\frac{1}{z^{-1-i(m-n)\omega}} = \frac{2\pi}{\omega}\delta_{mn}. \tag{33}$$

The quantum states then define an orthonormal set.

The dilatation operator of the conformal group defines supersymmetric Hamiltonians that annul the vacuum commensurate with the zeros of the Riemann ζ-function. For a quantum system with N quantum bits, each equiprobable $P_i = \rho_{ii} = \frac{1}{2}$, then entropy $S = -Tr[\rho log(\rho)]$ is $S = Nlog(2)$. The upper bound on complexity is $C_{max} = exp\,(N)$ [14]. A 10 solar mass black hole then has 10^{40} Planck mass units or qubits, and thus has entropy proportional to this. The time for a system to reach entropy or equilibrium is proportional to $ln\,(N)$, while the time it takes to reach maximum complexity is proportional to e^N. This is an incredibly long time period, say for stellar mass black hole $e^{e^{4\pi m^2}}T_p \simeq 10^{10^{10^{75}}}$ Planck times. This is comparable to the possible stability of a de Sitter vacuum.

The logistic function, which approximates a curve for the zeros of the Riemann ζ-function, depends on the number of states N in the entanglement or the N-tangle. If one adds a qubit this new state entangles with all these states doubling the complexity. The measure of entanglement is ultimately its complexity, which can be difficult to assess. In this case here with vacuum states, the number of vacuum states in this N-tangle defines complexity that is enormous. A recent paper [15] set bounds on what an entanglement witness can measure. The detectability of entanglement becomes exponentially difficult. Complexity increases as $C \sim e^S$ for $S = -Tr[\rho log(\rho)]$ such that if $\rho_{nn} \simeq e^{-n\beta}$, n the ordinal number of a qubit with probability $p_n = e^{-n\beta}$ and $\sum n = N$, then

$$S \simeq \sum_n n\beta e^{-n\beta} = \beta\sum_n \frac{\partial}{\partial\beta}e^{-n\beta} = N\beta. \tag{34}$$

The complexity then grows as $C \simeq e^N$. This also infers how small errors in determining these states, such as with state preparation form $S+\delta S$, errors in complexity can grow exponentially.

This doubling of complexity with the introduction of a single additional qubit illustrates how a noisy environment can introduce errors in complexity that grows enormously. For the system with d dimension and a reservoir of state with dimension k the detection capacity of some matrix M of observables is

$$C(M) = [Tr(M\rho) < 0], for\ \rho \simeq p_{d,k} = state\ distribution. \tag{35}$$

The entanglement witness fidelity is then $exp\left[-\frac{d-1}{k}/2(d+1)\right] \simeq e^{k/2}$, and the detectability is

$$C(M) < 2n\,exp\left[-\left(\sqrt{1+\frac{min}{TrM}\sqrt{tr(M)^2}-1)^2 k}\right)\right] \leq 2Ne^{-(3+2\sqrt{2})k}. \tag{36}$$

Hence, by building the complexity of a black hole by introducing more qubits will introduce more error. These qubits can just be polarizations of vacuum states with different momentum values. It is not possible to prevent the vacuum from entering a black hole, and the set of possible vacuum states may be impossible to control. For the system with dimension d, and reservoir of states with dimension k then $k < constd^{3/4}$, and the detection capacity is approximately $D(M) \simeq C^{-3/4}$.

An EPR pair of states, where on pair enters a black hole, will have its entanglement conserved — in principle. However, Bob outside will no longer be able to determine whether his state is entangled with the other pair. That other state becomes spread across the stretched horizon. Bob says his EPR pair is entangled with the black hole, but this is an entanglement between a quantum state and what is often thought of as a classical object. This is certainly the case for a multi-solar mass black hole. Similarly, a superposed quantum state that approaches a black hole will have its superposition lost. We may think of the black hole as a needle state so the quantum phase of the superposition is transferred to the black hole. A needle state can be a spin that detects a photon at a slit in a 2-slit experiment. The superposition of the photon state is converted into an entanglement with those states and the needle state. The needle state is then coupled to more quantum states and degrees of freedom, where the outcome is made classical. This is the collapse, or in the MWI the subjective perception of one branch in a global quantum system. The black hole serves the same role, where there is a phenomenological perception of the absorption of the quantum phase.

Quantum states in the presence of spacetimes with a Killing horizon, spacetimes with a Killing vector field orthogonal to a null surface, exhibit decoherence. The wave function associated with these states has some "tail" that crosses this surface and there is no causal connection to send a signal to the exterior. For this reason Alice inside a black hole cannot teleport a state to Bob in the exterior. Bob is entangled with Alice's quantum state on the stretched horizon, but there is no entanglement with the EPR pair after it passes the horizon on its proper time. This has been shown to an inescapable aspect of how a quantum state behaves with a black hole [16]. This is in part commensurate with the above discussion in the limit that the number of states is infinite or in the limit $\lim_{d\to\infty}$ complexity diverges. This is the case where quantum complexity is determined by the zeros of the Riemann ζ-function. The zeros of the Riemann ζ-function serve as Godel number for the set of quantum states or vacuum states of the black hole.

The detection capacity is absolutely zero in the infinite case, and where the prime distribution is a sort of Godel numbering scheme for the vacuum states which compose it. Prime factorization is not NP-complete for it is in $NP \cap co - NP$, though whether it is not P is unclear. Prime factorization is in NP, because a decision whether n has a prime factor p with k bound $p < k$ then $p|n$ serves as a witness of a yes instance. Prime factorization is in $co - NP$ because a prime factorization of n with no factors less than k serves as a witness of a no instance. Prime factorizations are unique and P, because testing for primality is in P. So far nobody has succeeded in proving that prime factorization is P [17], and as yet whether $P = NP$ is an open question.

5. Schwarzian and supersymmetry

Quantum mechanics is based on eigenvalued system of state vectors and operator. For the operator a second order differential a typical Sturm-Liouville form of differential equation is

$$\psi_{xx} + \phi(x)\psi = 0. \tag{37}$$

Here we think of the $i\partial\psi/\partial t = \phi(x)\psi$ in the Schrödinger equation. The Wronskian determinant defines two solutions ψ_1 and ψ_2 so the quotient $f = \phi_1/\phi_2$. the potential function $\phi(x)$ has the form $\phi(x) = \frac{1}{2}S(f)$ [18]. This means the Schwarzian is a form of Hamiltonian. This quotient is an affine coordinate on the projective line $\mathbb{R}P^1$ so that the Hamiltonian is a parameterized curve on this projective line. For the metric $g = dxdy$ and $y = u(x)$ the curvature is $R = u_{xx}/(u_x)^{3/2} = S(u)/\sqrt{u}$, which is the Schwarzian derivative and its connection to curvature. This curvature is associated with hyperbolic dynamics. The geodesic flow on a hyperbolic space is governed by the curvature $R < 0$.

The Virasoro algebra is defined with a basis $L_m = e^{im\theta}\partial_\theta$, where it is not hard to see the commutator $[L^m, L^n] = i(n-m)L^{m+n}$ holds. This is the Witt algebra that is extended to the Virasoro algebra. The state operators (L_{-1}, L_0, L_1) form a closed algebra $SL(2,\mathbb{R}) \simeq SU(1,1)$. This is a system of boosts in special relativity. This means the system has an invariant given by a function $\psi(x)$ such that the Schwarzian

$$S(\psi) = \psi_{xxx}/\psi_x - (3/2)(\psi_{xx}/\psi_x)^2 \tag{38}$$

is invariant $S(\psi') = S(\psi)$ under $SL(2,\mathbb{R})$ transformations

$$\psi' = \frac{a\psi + b}{c\psi + d} \tag{39}$$

This function $\psi(x)$ corresponds to the states (L_{-1}, L_0, L_1) with the $SL(2,\mathbb{R})$ transformations of this function

$$\psi' = a\psi + b, \psi' = a\psi + b\psi, \psi' = \psi + c\psi^2 \tag{40}$$

define the elementary generators

$$P = i\partial_\psi.D = i\psi\partial_\psi, K = i\psi^2\partial_\psi, \tag{41}$$

For translation, dilatation, and special conformal transformation acting upon the form of the field $\psi(x)$. This closed algebra has bounded orbit space and there its eigenspectrum is bounded below. This is despite the negative curvature of hyperbolic space and the negative vacuum that implies.

We then have a connection between the Schwarzian and the Hamiltonian. The Schwarzian form of the Hamiltonian bridges the linear quantum Hamiltonian with that of nonlinear dynamics seen in classical mechanics. Such nonlinear dynamics is often solitonic. It is interesting to note that the Korteweg-deVries (KdV) equation [19] is of the form

$$u_{xxx} - 6uu_x = u_t, \tag{42}$$

which is similar to the Schwarzian. This is an equation for the solitary wave observed in a canal with the solution

$$u(x) = \frac{1}{2}vsech^2\left[\frac{1}{2}\sqrt{v}(x - vt + \delta)\right]. \tag{43}$$

The Lax pair is a commutator that describes the dynamics according to

$$L_t = [L, T], \tag{44}$$

which for $L = -\partial_x^2 + u$ and $T = 4\partial_x^3 - (2u\partial_x + \partial_x u)$ defines $L_t = u_t$ in the KdV equation. Clearly L is a Sturm-Liouville operator, and we can then inquire whether the KdV equation is a form of the Schwarzian. In particular u_{xxx}/u_x is suggestive of a Schwarzian form of the KdV equation. To address this we move to consider an additional term $T' = -6u\partial_x + v$ so that

$$[L, T'] = \partial_x^2 v + 6u\partial_x u \tag{45}$$

Then for the entire $T'' = T + T'$ the Lax Pair commutator gives

$$\left[L, T''\right] = u_{xxx} + v_{xx} = u_t, \tag{46}$$

where $v_{xx} = -(3/2)u_{xx}^2/u_x$. This equation is

$$\left[u_{xxx}/u_x - (3/2)(u_{xx}/u_x)^2\right]u_x = u_t, \tag{47}$$

or $S(u)u_x = u_t$. To pursue this further requires a Bäcklund transformation and Pfaff forms, similar to the Estabrook-Walquist method for the KdV [20].

This KdV solution may be extended into supersymmetry with

$$\Phi(x, \theta) = \phi(x) + \theta\psi(x), \tag{48}$$

in such a way that $\phi\phi_x \to D\Phi\Phi_x$, for $D = \partial/\partial\theta + \theta\partial/\partial x$ corresponding to the supergenerator $Q = \partial/\partial\theta - \theta\partial/\partial x$. The SUSY KdV is then

$$\Phi_{xxx} + (D\Phi)\Phi_x + \Phi_t = 0. \tag{49}$$

With supersymmetry $D\Phi = 0$ and $\Phi_{xxx} + \Phi_t = 0$, or $\Phi_{xxx}/\Phi_x + \Phi_t/\Phi_x = 0$, and we can see this is $S(\Phi)\Phi = \Phi_t$. The extension with the Lax pair may be then compared to a Bäcklund transformation with $D\Psi = \Phi$ so that

$$\Psi_{xxx} + \frac{1}{2}\Psi_x^2 + \Psi_t = 0. \tag{50}$$

With $\Psi(x,\theta) = \chi(x) + \theta\omega(x)D\Psi = \Phi$ means $\phi = \partial_x B$ and $\omega = \phi$. We then have

$$\begin{aligned} \phi_{xxx} + \phi\phi_x + \phi_t &= 0 \\ \Psi_{xxx} + \phi\psi_x + \psi_x &= 0. \end{aligned} \tag{51}$$

This little analysis is interesting in that it points to a prospect for understanding how quantum information may be converted into spacetime information. The differential equation

$$\phi_{xx} + \frac{1}{2}u(x)\phi = 0 \tag{52}$$

may well be a Schrödinger equation with $i\partial_t\phi = \frac{1}{2}u(x)\phi$, so we may write

$$\phi(x,t) = \phi(x)e^{-i\int u(x,t)dt} \tag{53}$$

So there is the nonlinear curvature $S(u) = u_{xxx}/u_x - \frac{3}{2}(u_{xx}/u_x)^2$, and this may be extended to soliton mechanics. Solitons are nonlinear waves that are not quantum mechanical. In this way quantum information may be converted into an entirely different form of information; the information is still conserved — which should hold for spacetimes with a timelike Killing isometry.

Quantum mechanics is strictly linear, while spacetime physics can be extremely nonlinear. For this reason, it has been difficult to derive a concept of quantum gravitation. The use of complex linear vector spaces, which derive eigenvalues for eigenfunctions is central to quantum theory. Even quantum field theory needs this, which is just a set of operators on an empty Fock space that define states. In fact, it is being suggested that quantum gravitation is not possible [21]. Yet with this elementary equivalency of Sturm-Liouville systems and the Schwarzian, and then further equivalences with soliton waves with supersymmetry, we may have that quantum mechanics and spacetime physics are ultimately different aspects of the same thing.

Topological quantum field theory is a categorical equivalency between manifolds of space and complex linear vector spaces [22]. The cobordism hypothesis with TQFT means that bordisms between manifolds are ultimately equivalent to transformations on complex linear vector spaces. These complex linear vector spaces may serve as quantum states and a finite Hilbert vector space. In this setting classical gravitation may in a veiled manner be quantized, where spacetime manifolds are categorically equivalent to quantum states. This is a subject this author is pursuing. At this time such detailed discussions on this must await a subsequent publication. However, recent developments by Jacobson, van Raamsdonk [23] and Susskind [24] indicate how spacetime is equivalent or built from quantum entanglements.

6. Quantum computing implications

We now consider how this can impact on the foundation of computation and practical computing. The quantum gravitation of black holes may appear far removed from practical computing. This physics may only at best be observed in the physics of gravitational waves produced by black hole coalescence [25]. However, there are analogues of black holes and Hawking radiation in quantum optical experiments [26].

These physical experiments may provide tools in future computing systems. If we are to consider quantum information as conserved, then black holes are ultimately forms of quantum computers. In fact Aaronson and Watrous have illustrated how quantum computers are equivalent to a computer with closed timelike curves [27]. Closed timelike curves are a feature of black hole interiors and the anti-de Sitter spacetime.

Computers rely upon physics to construct electronic systems. Solid state physics is a central aspect of quantum electronics. Since this paper began with the subject of supersymmetry, consider how supersymmetry may exist as an emergent system in solid state physics. This may in fact have some connection with anyonic physics.

Supersymmetry is the intertwining between a boson ϕ and a fermion ψ is

$$\delta\phi = \bar{\xi}\psi \tag{54}$$

$$\delta\psi = i\bar{\xi}\sigma^{\mu}\partial_{\mu}\phi, \tag{55}$$

for ξ an infinitesimal Grassmann variable ([28], p. 22). The standard actions for the boson and fermion summed are

$$S = -\int dtd^{n}x\left[|\partial_t\phi|^2 - \left|v_b\partial_j\phi\right|^2 + i\overline{\psi}\left(\partial_t + v_f\sigma_{\mu}\partial_{\mu}\right)\psi\right], \tag{56}$$

where $n = 3$. The supersymmetric transformed action is

$$\delta S = \int dtd^{n}x\left(\xi\partial_{\mu}(\overline{\psi}\partial_{\nu}\phi) + \bar{\xi}\partial_{\mu}(\partial_{\nu}\phi^{*})\psi\right). \tag{57}$$

This is an overall boundary term and is zero. This action is invariant under SUSY transformations. The Lagrangian density defines the Hamiltonian density $\mathcal{H}$ with $\mathcal{L} = \left|\dot{\phi}\right|^2 - i\overline{\psi}\psi - \mathcal{H}$ so that

$$\mathcal{H} = \left|\partial_j\phi\right|^2 - i\overline{\psi}\sigma^{i}\partial_i\psi. \tag{58}$$

The supergenerator, which previously expressed according to $a^{\dagger}b$ is considered with the boson and fermion fields ϕ and ψ as

$$Q = \int d^2x\overline{\psi}\sigma^{j}\partial_j\phi. \tag{59}$$

The anticommutator $\{Q, \overline{Q}\}$ with the Fourier expansion in box normalization $\phi(x) = L^{-1/2}\sum_k\phi_k e^{i\mathbf{k}\cdot\mathbf{x}}$, and $\psi(x) = L^{-1/2}\sum_k\psi_k e^{i\mathbf{k}\cdot\mathbf{x}}$, gives

$$\{Q, \overline{Q}\} = \frac{1}{L}\sum_k\left(k^2|\phi_k|^2 + k|\psi_k|^2\right). \tag{60}$$

which is the standard Hamiltonian for bosons and fermions. In the evaluation of the anticommutator there is the application of the operator $\sigma^{j}\partial_j$, where with the meaning in a total time derivative $d/dt = \partial_t + v_j\partial_j$. Since the derivative is on the bosons this means the velocity is the velocity of the bosons. However, this Hamiltonian is correct for equal velocity of fermions and bosons $v_f = v_b$. If this condition is not satisfied, there is then an additional term in $v_b - v_f$ that represents the breaking of supersymmetry [29]. It is also associated with a deviation from Majorana statistics.

The scalar superfield $\Phi = \phi(x) + \sqrt{2}\theta\psi(x) + \sqrt{2}\overline{\theta}\overline{\psi}(x)$ the theory uses the supercharge

$$Q = -i\frac{\partial}{\partial\theta} + \overline{\theta}\sigma^{\mu}\frac{\partial}{\partial x^{\mu}} \tag{61}$$

and the superderivative

$$D = \frac{\partial}{\partial\theta} - i\overline{\theta}\sigma^{\mu}\frac{\partial}{\partial x^{\mu}}. \tag{62}$$

The anticommutators $\{Q, \overline{Q}\} = \{D, \overline{D}\} = 2i\sigma^{\mu}\partial_{x}^{\mu}$, which is the standard SUSY result, the anticommutation of supercharges is a generator of a Lorentz boost. The Lagrangian with terms $W^{a}W_{a}|_{\theta\theta} + W^{\dot{a}}W_{\dot{a}}|_{\overline{\theta\theta}}$ applies to the bulk field. Edge state require a Lagrangian density with terms $\phi^{*}(x)\phi(x + v\partial_{x})$ and $\psi^{*}(x)\psi(x + v\partial_{x})$. The corresponding supersymmetric action is

$$S = \int dtd^{2}x\left(\int d\theta\overline{\Phi}(x)\Phi(x + v\partial_{x})\right) + \int d\overline{\theta}\overline{\Phi}(x + v\partial_{x})\Phi(x)). \tag{63}$$

We then have $\Phi(x + v\partial_{x}) = \Phi(x) + \epsilon D\Phi(x)_{\theta}$ and the action is

$$S = \int dtd^{2}x\left(\int d\theta\left(\overline{\Phi}(x)\left[\Phi(x) + v\epsilon D\Phi(x)_{\theta}\right]\right)\right) + \int d\overline{\theta}\left(\overline{\Phi}(x) + v\epsilon\overline{D\Phi}(x)_{\overline{\theta}}\right])\Phi(x). \tag{64}$$

The $x \to x + v\partial_{x}$ in the supersymmetric setting is determined by the group element $g(x, \theta, \overline{\theta}) = e^{-i\left(x^{\mu}p_{\mu} + \theta\overline{Q} + \overline{\theta}Q\right)}$. The reparameterization of a configuration variable $x^{\mu} \to x^{\mu} + i\theta\sigma^{\mu}\overline{\xi} - i\overline{\theta}\sigma^{\mu}\xi$, is accomplished with the composition $g(0, \xi, \overline{\xi})g(x, \theta, \overline{\theta})$. The $\Phi(x) \to \Phi(x) + v\epsilon D\Phi(x)_{\theta}$, $\epsilon > 0$ a small number, is a rapidity which is a supersymmetric transformation.

Supersymmetry is now considered in the context of a periodic system, such as a lattice or crystal. Before supersymmetry we first just consider a gauge symmetry. We may consider this in a form of Bloch wave [30], where a gauge potential is transformed by the action of a periodic unitary operator. The Bloch wave function $\psi(x) = e^{i\vec{k}\cdot\vec{x}}u_{k}(x)$ in the Schrodinger equation results in

$$\frac{\hbar^{2}}{2m}\left(-i\nabla + \vec{k}\right)^{2}u_{k}(x) + V(x)u_{k}(x) = Eu_{k}(x) \tag{65}$$

whith an effective momentum $P = -i\hbar\left(\nabla + i\vec{k}\right)$. This lattice momentum is a sympletic transformation of the momentum, or with the $\vec{k} = -i\vec{A}$. This leads to the prospect for $\nabla\cdot\vec{A}$ terms, but these may be removed with the Coulomb gauge condition. This does indicate the lattice induced $e^{i\vec{k}\cdot\vec{x}}$ can be generalized into a unitary group transformation. Such a general transformation of a phase the potential $\vec{A}$ occurs on the unitary matrix $e^{-i\vec{A}\cdot\mathbf{e}} = e^{-i\alpha}$ as

$$U_\phi^\dagger e^{-i\phi} U_\phi = e^{i\vec{k}\cdot\vec{x}} e^{-i\alpha'}. \tag{66}$$

This is a periodic gauge symmetry condition [31]. The units $\mathbf{e}_i$ are such that $\mathbf{e}_i\mathbf{e}_j = \frac{1}{2}[\mathbf{e}_i, \mathbf{e}_j]$, which makes these effectively one-forms. The transformation of a gauge potential from lattice site to lattice site induces a Bloch wave structure on the transformed potential. The standard technique of letting $U_\phi = e^{-i\phi} \simeq 1 - i\phi$ and $e^{-i\alpha} \simeq 1 - i\alpha$ results in

$$\alpha' = \alpha + ie^{i\vec{k}\cdot\vec{x}}[\alpha, \phi] - e^{\frac{i}{2}\vec{k}\cdot\vec{x}} sin\left(\frac{i}{2}\vec{k}\cdot\vec{x}\right). \tag{67}$$

The operator $\alpha = e^i\partial_i$ gives $[\alpha, \phi] = e^i\partial_i\phi$, which corresponds to a pure gauge term. Now generalize U_ϕ to $G(0, \theta, \overline{\theta})$ as a supergroup element

$$G(0, \theta, \overline{\theta}) = e^{-i(\theta Q + \overline{\theta Q})}. \tag{68}$$

The momentum operator in general is the covariant $P_\mu = p_\mu + ieA_\mu$. Consider $\alpha = x^\mu P_\mu - \xi Q - \overline{\xi Q}$ and product $G(0, \xi, \overline{\xi})G(x, \theta, \overline{\theta})$ is then

$$G(0, \xi, \overline{\xi})G(x, \theta, \overline{\theta}) = exp\left[i\left(x^\mu P_\mu + i\theta\sigma_\mu\overline{\xi} + i\xi\sigma_\mu\overline{\theta}, \theta + \xi, \overline{\theta} + \overline{\xi}\right)\right]. \tag{69}$$

The transformation

$$G(0, \xi, \overline{\xi})G(x, \theta, \overline{\theta}) = e^{ik_\mu x^\mu} G'(x, \theta, \overline{\theta})G(0, \xi, \overline{\xi}), \tag{70}$$

defines $G'(x, \theta, \overline{\theta})$ with $P'_\mu = p_\mu + A'_\mu$, a gauge transformed gauge potential. The transformed gauge potential is $e^{ikx}A_{\mu'} = A_\mu + \delta A_\mu$, for $\delta A_\mu = C_\mu(\xi + \theta + \overline{\xi} + \overline{\theta})$. Here C_μ comes from the commutator $[Q, A_\mu]$. The Grassmann elements then define the gauge transformation. The differential operator Q is then $Q_\alpha = \partial_{\theta^\alpha} - i\sigma^\mu_{\alpha\alpha'}\overline{\theta}^{\alpha'}D_\mu$ and the action of the Grassmannians $\xi, \overline{\xi}$ transforms the gauge potential in a way analogous to $A' = gAg^{-1} - i\partial g g^{-1}$. It is important also to note the Bloch wave factor so the transformation is in total

$$A'_\mu = e^{-ikx}\left[A_\mu + C_\mu(\xi + \theta + \overline{\xi} + \overline{\theta})\right]. \tag{71}$$

This is then a lattice approach to supersymmetry, where we may replace the e^{ikx} with $e^{ikx} \to e^{ikx_n}$ for lattice sites labeled by n, and further the potentials and Grassmann variables exist at lattice sites as well. These transformations occur per lattice site and independent of each other consistent with Wightman conditions and axioms of quantum field theory.

Supersymmetry is a basis for quantum gravitation, and it was proposed with the interest in possible connections between spacetime physics and quantum physics. As stated above, general relativity as a classical field theory is nonlinear and its gauge-like field is extrinsic. This means the wave function of gravitation is also nonlinear and the linear rules of quantum mechanics do not apply.

A standard case is the nonlinear Schrodinger equation

$$i\frac{\partial\psi}{\partial t} = -\alpha\frac{\partial^2\psi}{\partial x^2} - \sigma|\psi|^2\psi. \tag{72}$$

The extrinsic nature of this can be seen where $\sigma|\psi|^2$ acts as the potential in the Schrodinger equation, but where this depends on the wave function. The solution to this equation is a soliton equation, the standard solution is a tanh function. If we ignore the nonlinear term $\sigma|\psi|^2\psi$ this is a simple Schrodinger wave equation for a free particle. We may then split the wave equation into its linear and nonlinear parts so that

$$i\frac{\partial\psi}{\partial t} = i\frac{\partial}{\partial t}(\psi_\ell + \psi_n) = (V_\ell + V_n)\psi, \tag{73}$$

where V_ℓ and $V_n = \sigma|\psi|^2$ are the linear and nonlinear potential functions. Perturbation methods clearly can be applied for $\omega \ll \alpha$.

The split set Fourier transform method. This can be seen with a polar form of the wave function $\psi(x,t) = \rho(x,t)e^{iS}$, so that $\rho(x,t)\partial(\rho x, t/\partial t) \ll i\partial(Sx, t/\partial t)$ Then consider the form of

$$\psi(x,t+\tau) = exp\left[i\sigma|\psi(x,t)|^2\tau\right]\psi(x,t) \tag{74}$$

where the time increment $\tau \ll t$. Then clearly

$$lim_{\tau\to 0}\frac{\partial}{\partial\tau}\psi(x,t+\tau) = \ |\psi(x,t)|^2\psi_n(x,t) \tag{75}$$

The small time τ incremented nonlinear portion will define a dispersion that is defined by the Fourier transform

$$\chi(k,t) = \int_{-\infty}^{\infty}\psi_n(x,t)e^{i(k-k_0)x}dx \tag{76}$$

so the small near linear time step defines

$$\chi(k,t+\tau) = \psi_n(x,t)e^{\frac{i\alpha}{2}(k-k_0)^2\tau}\chi_n(k,t) \tag{77}$$

We then perform the inverse Fourier transform

$$\psi(x,t+\tau) = \frac{1}{2\pi}\int_{-\infty}^{\infty}e^{-ikx}\chi(k,t+\tau)dk. \tag{78}$$

By performing this N times for small time increments The soliton may be advanced in time and thus in space. This method can be applied to many soliton equations.

This process indicates that the physics of quantum mechanics, incrementally stepped in time, still pertains. The linear part is analytic by linear algebraic techniques and can be called quantum mechanical. The nonlinear term is not, but with the small step procedure the errors form this term can be made relatively small. The pulse then conserves quantum information. We may then make the statement that nonlinear physics, such as solitons, will conserve information, indeed quantum information, but in a different form.

Black holes as quantum computers may stand as the ultimate form of computation. What limits there may exist in computation or computability may occur in black holes. This may also involve questions on the conservation of quantum information in black holes. The Kerr solution for instance has an inner horizon that is continuous with $\mathcal{I}^+$, which means that an infinite number of light pulses and equivalently information can enter. The observer who proceeds across this horizon may in fact be able to perform hypercomputation, such as with Malement-Hogarth spacetimes [32], and this is a loop around the Gödel-Turing limit on computation. The occurrence of Hawking radiation however may break this system. The laws of physics may then in effect enforce the Church-Turing thesis of computation [33].

This brings in the physics of Hawking-Page temperature and the Page time. It has been my conjecture that there is a phase transition in the properties of a black hole at different temperatures $T = 1/8\pi m$. I conjecture that there is something real about the so-called firewall, in that once a black hole has quantum evaporated to some state it assumes different properties. Susskind and Lindesay in their book discuss how a black hole is a conductor. However, maybe at a low enough temperature it becomes a Mott insulator, where in effect the black hole is now sealed off and interacts with the outside world only with its exterior gravity. This may be a limitation inherent in physics to protect the Church-Turing thesis.

The breakdown in the $ER = EPR$ or the equivalency between quantum states CFT_3 and gravity in AdS_4, here I consider as the above correspondence between Kerr interior and AdS_4, is not about the destruction of quantum information. The topology of AdS_n is $\mathbb{R}^{n-1} \times \mathbb{S}^1$. The cohomological groups for AdS_n should be

$$H^0(AdS) = \mathbb{R} \tag{79}$$

$$H^1(AdS) = \mathbb{R} \tag{80}$$

$$H^{n-1}(AdS) = 0 \tag{81}$$

$$H^n(AdS) = 0 \tag{82}$$

The topology of a Schwarzschild black hole is $\mathbb{R}^2 \times \mathbb{S}^2$ which has $H^0(M) = \mathbb{R}$, $H^2(M) = \mathbb{R}$, and all else are zero. The Kerr metric has topology $\mathbb{S}^2 \times \mathbb{R}^1 \times \mathbb{R}^1$ in its interior. The transformation of $\mathbb{R}^2 \to \mathbb{R}^1 \times \mathbb{S}^1$ indicates a singularity in the interior. This then provides $H^1(M) = \mathbb{R}^1$. Further, if this restricted to the inner ergosphere, a region that bounds closed timelike curves, we may then consider $H^1(M) = 0$. Thus, for the region of interest the Kerr interior has the same, or nearly so, topology as AdS_4. The inner ergosphere defines a region that is a trapping region, so curves that enter never leave. Thus, quantum information should not be destroyed. It is interesting to note that this interior region with closed timelike curves connects with Aaronson an Watrous on the relationship between closed timelike curves and quantum computation [27].

This issue may concern computational incompleteness in any possible quantum error correction algorithm in the dynamics of the black hole. Consider an error correction program that is capable of assuring that any quantum bit of information (qubit) that is sent into the Hawking wedge or island [34] as free radiation registers a positive. This is then a register system that records a unit out of the alphabet $\{0, 1\}$. If the register sets a 0 that is a failure. The goal is a maximal set of 1's. This means the extremization of the error correction is a Busy Beaver problem, which Tibor Radó showed is not generally computable. The quantum information that escapes is not so much destroyed as it is not accessible.

There is a connection to the foundation of quantum mechanics. Hossenfelder and Palmer have assumed that statistical independence is false, and this is a loop-hole around the Bell inequalities. This leads to invariant set theory for QM, which has p-adic context. There is also an incomputable nature to this by the Matiyasevich theorem [5]. Also, from a basic physics perspective the loss of statistical independence means there is an infinite fine tuning to quantum states that may be incomputable in the Busy-Beaver sense. I think their super-determinism is wrong, but as Wheeler once said, the falsity of a good hypothesis can itself be a great result.

Black hole horizons in 4-dimensions have spherical symmetry or $\mathbb{S}^2$. In 5-dimensions the horizon can have symmetry $\mathbb{S}^3$, $\mathbb{S}^2 \times \mathbb{S}^1$ or the Lens space $L(q,3) \simeq \mathbb{S}^3/\mathbb{Z}_p$ [35]. The last is a topology that equates points on the sphere with a period lattice. Solutions in 5-dimensions have $U(1)^2$ rotational symmetry. In this setting of bi-axial-symmetry, the list of possible horizon topologies reduces to the sphere $\mathbb{S}^3$, the ring $\mathbb{S}^1 \times \mathbb{S}^2$, and the lens spaces $L(p,3) \simeq \mathbb{S}^3/\mathbb{Z}_p$, for any choice of relatively prime integers $1 \leq q < p$. It is also possible to classify the list of possible domain of outer communication (DOC) topologies in this regime. Namely, the compactified Cauchy surfaces within the DOC must either be the 4-sphere $\mathbb{S}^4$, a connected sum of $\mathbb{S}^2 \times \mathbb{S}^2$, or in the non-spin case a connected sum of complex projective planes $\mathbb{C}P^2$ and $\mathbb{C}P^2$. There have been a number of attempts to realize all the topologies in these lists, however they have all suffered from the presence of either naked singularities, or conical singularities on the axes, when trying to implement the more complicated configurations.

This leads to future work that will tie the foundation of mathematics and computation to physics, in particular physics that may define computational platforms. The work on $P = NP$ or its opposite has lead to curious possible connections between geometric complexity, the Riemann ζ-function and $P = NP$. There may in fact exist some higher order Riemann ζ-function [36]. The role of supersymmetry is the structure of the vacuum may play a central role. There may in fact be deep considerations with computability with the Langlands program and it connection to the foundations of mathematics and theoretical physics.

The practical structure of computers and algorithms themselves will likely then assume properties of physical theories, in particular quantum mechanics. Optical synthetic black hole are laboratories for exploring the limits of computability. Further into the future architectures of this sort may be integrated into standard technology.

7. Systems and simulations

The Lie group $SL(N,\mathbb{R})$ has dimension $d = N^2 - 1$. The simplest of for $N = 2$ which is equal to $SU(1,1)$. The Pauli matrices for $SU(2)$ are σ_x, σ_y and σ_z. The $SU(1,1)$ can be obtained with the replacement of the Pauli matrices with $\tau = i\sigma_i$, which carries signatures of the old *ict* approach to relativity. This has three dimensions, and the geometry of a surface of constraint is not a sphere, such as a Bloch sphere, but a hyperboloid. These are the system of boosts in special relativity. The Lie group $SL(2,\mathbb{C}) = SU(1,1) \times SU(2)$ may be thought of as the boosts plus 3-rotations. A spinor representation is found in the Dirac matrices. In fact, with $\{\gamma^\mu, \gamma^\nu\} = -2g^{\mu\nu}$ The Dirac equation written as $i\partial_m u\gamma^\nu\psi = m\psi$ defines the connection so

$$i\partial_m u\gamma^\mu\psi = \left[i\gamma^\mu\partial_\mu + \Gamma^\mu_{\mu\nu}\gamma^\nu\right]\psi, \tag{83}$$

which could be written as a covariant derivative on the Dirac field. The Dirac operator $D = i\partial_m u\gamma^\mu$, setting the mass to zero, is equivalent to the field $D \sim \psi$, and so this is the same as $D^2 = 0$, which is in differential forms $d\Gamma + \Gamma\wedge\Gamma = 0$. If we consider the mass, then we have

$$D^2 \sim D\psi = imD. \tag{84}$$

Now $m^2 \sim D^2$ and $\overline{\psi}D^2\psi$ is equivalent to $\overline{\psi\psi'}\psi\psi'$ as a quartic Lagrangian, where the mass assumes a role similar to curvature. The Dirac operator then functionally builds $\mathcal{R} = d\Gamma + \Gamma\wedge\Gamma$.

We may then functionally build gravitation as $SL(2,\mathbb{C}) = SU(1,1) \times SU(2)$ as entangled spins, where one spin is has $\sigma_z \rightarrow \tau_z\ i\sigma_z$. This may be emulated for one of the spin in a tunneling barrier or a region where the kinetic energy $\hbar^2k^2/2m = \sqrt{E-V}$ for $V > E$. This could be emulated in a quantum dot where an electron is in the trapped material of the quantum dot. Similarly, for $SL(3,\mathbb{C})$ this is 16 dimensional and effectively the product $SU(3) \times SU(2,1)$. Is the complex-complex $\mathbb{C}/\mathbb{C}$ or $A_2 \times A_2$ on the Freudenthal-Tits magic square leading up to the octonions. This may be extended further to $SL(4,\mathbb{C})$ is 32 dimensions. This is equivalent to $U(4) \times U(2,2)$ or $SO(6) \times SO(4,2)$, which is $\sim SO(6,\mathbb{C})$. This is not on the Freudenthal-Tits magic square, but this has the same dimension as $\mathbb{O}\mathbf{P}^2$ or the dimension of the dual to the Fano plane $\mathbb{O}\mathbf{P}$. In this system also lies the $AdS_5 = SO(4,2)/SO(4,1)$, such that $SO(4,2) = SU(4)$ is the isometry group. This is a further build towards modeling physics.

This has considered the tunneling or instanton state. The occurrence of a spin in a region with a potential larger than the energy will model the Lorentzian space by this imaginary time. An electron in a quantum dot is a trapping region and in the limit the dot is a point the trapping approaches a delta function potential. The wave function is then $\sim e^{-k|x|}$ from the dot at the origin. For a finite potential region there are complex wave functions. We can then extend this physical model from group theoretic considerations to modeling black holes.

We consider time-loops or CTCs in relationship to entanglements. Further, this is done in the framework of a pseudo-black hole. Quantum states interior and exterior are entangled where the entanglement of the exterior state with another state is equivalent to the evolution of the interior state into the corresponding that that exterior state is entangled with.

Time loops in a quantum mechanical setting must have a measure of constructive interference. If they completely destructively interfere then there is no loop. CTCs are not physically likely in the open world. It requires a certain type of spacetime for these to occur, such as the anti-de Sitter spacetime or the timelike interior of a Kerr-Newman black hole. This is not exactly the spacetime we experience. Consider quantum fields in the timelike interior with CTCs, such as the timelike interior of a Kerr-Newman (KN) black hole. This quantum wave has some extent, and it is red-shifted form leaks out of the black hole. You then entangle your quantum wave with this. The quantum wave you are working on is selected, say by Hadamard gate process, which by entanglement selects for the interior quantum state. You now can perform quantum gravitation computing where your line directed quantum computer or quantum states are entangled with quantum states on CTCs in the black hole interior (**Figure 1**).

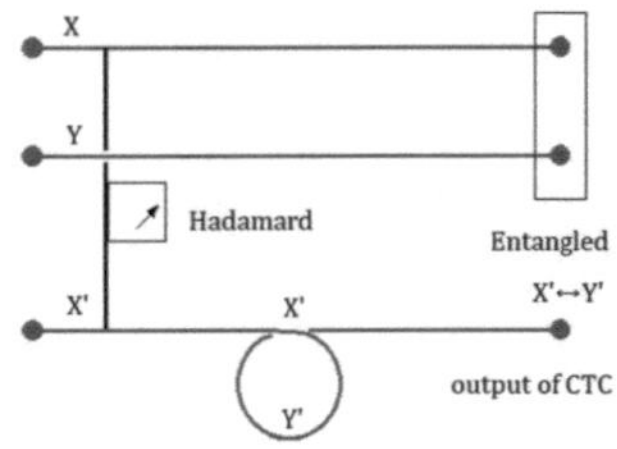

Figure 1.
Entanglement and interior oscillation as a form of entanglement swap.

The diagram here illustrates this situation. On the left we have the entanglement of two states, represented by $x \oplus y$. We may think of this evolution as a Hadamard gate on x that in a bipartite setting evolves $|x\rangle \to \frac{1}{\sqrt{2}}(|x\rangle \pm |y\rangle$, or even more explicitly

$$|1\rangle_x|0\rangle = \ |1\rangle|0\rangle \to \frac{1}{\sqrt{2}}(|1\rangle|0\rangle \ \pm \ |0\rangle|1\rangle) \tag{85}$$

This entanglement on the left is then equivalent to a situation in a timelike interior of a KN black hole. This means in the interior we have a unitary evolution $U|x\rangle = \ |y\rangle$ followed by $U'|y\rangle = x\rangle$. We must then have $UU' =$ or that $U' = U^{-1} = U^{\dagger}$. We may also think of this with $U = e^{-iHt/\hbar}$ and that the time reverse is then just the inverse of this unitary operator or equivalently the Hermitian conjugate $U = U^{-1} = e^{iHt/\hbar} = U^{\dagger}$. The loop will then forward evolve a quantum state and then reverse it, and in a loop this will continue.

We consider the cyclic time evolution of the interior states $|x\rangle \to |y\rangle \to |x\rangle$ according to a unitary evolution $U|x\rangle = e^{i\phi}|x\rangle$ or

$$e^{i\phi}|x\rangle = cos\phi|x\rangle + isin\phi|y\rangle. \tag{86}$$

Next consider an entanglement of the form $U(t)|x\rangle \otimes |m\rangle$ with the medium states $\{e^{-i\omega^+t}|+\rangle, e^{-i\omega^-t}|-\rangle\}$. This is then an entanglement of the form

$$U(t)|x\rangle \otimes |m\rangle = e^{-i\omega^+t}cos\phi|x\rangle|+\rangle + ie^{-i\omega^-t}sin\phi|y\rangle|-\rangle \tag{87}$$

This is expanded into

$$U(t)|x\rangle \otimes |m\rangle = \ cos\,(\omega^+t)\,cos\,(\phi)|1\rangle_x|+\rangle \ + \ sin\,(\omega^-t)\,sin\,(\phi)|0\rangle_x|-\rangle \tag{88}$$

We now perform an entanglement swap, where the superposition between $|x\rangle$ and $|y\rangle$ and the entanglement with the medium states $|m^{\pm}\rangle$ is converted to a superposition. The two states of $|x\rangle$ and $|y\rangle$ are sent to the proverbial Alice and Bob, and the environment state is projected by Hadamard gate operation onto them in a manner outlined by Coecke [37]. This gives the entangled quantum state

$$|x\rangle \oplus |y\rangle \to e^{-i\omega^+t}cos\phi|x\rangle|+\rangle + ie^{-i\omega^-t}sin\phi|y\rangle|-\rangle. \tag{89}$$

The time direction sensitive parts are $sin\,(\omega^-t)\,sin\,(\phi)|y\rangle|-\rangle$ and $isin(\omega^+t)\,cos\,(\phi)|x\rangle|+\rangle$. These mover relative to the other terms with $cos\,(\omega^{\pm}t$, which emulates a time loop. Thus it is possible to perform an entanglement swap on quantum states in a sort of synthetic time loop, similar to Rabi oscillation, so these states become entangled.

The above entanglement occurs in a black hole with a spacelike region that acts as a tunneling barrier. It is possible to model this with the standard tunneling barrier problem with quantum mechanics. In the case of the quantum optical analogue of the black hole, the barrier is modeled as moving in one direction. We then break the symmetry of the reflection and transmission coefficients. The interior of the black hole is modeled as the right region, the spacelike trapping region the potential well between $x \in (0, a)$. The particle in the interior moves to the right and enters the region highly red-shifted relative to the barrier. Conversely on the left particles enter the barrier highly blue shifted. We ignore such incident particles We then impose the condition that $k_\ell \gg k_r$ We then have the condition that $e^{iak_r} \simeq 1$. The diagram illustrates the wave amplitudes, with $A_{r,\ell}$ t$B_{r,\ell}$ $C_{r,\ell}$ the right waves, waves in the barrier, and left waves respectively, matching these boundary conditions are

$$A_r + A_\ell = B_r + B_\ell \tag{90}$$

$$k_0 A_\ell = k_1 B_\ell \tag{91}$$

$$B_r + B_\ell e^{-iak_\ell} = C_r + C_\ell e^{iak_\ell} \tag{92}$$

$$k_\ell B_\ell e^{iak_\ell} = k_\ell B_\ell e^{iak_0}. \tag{93}$$

The wave vector k_ℓ is designated now as just k, k_r is ignored for simplicity, and $k_0 = \sqrt{E - V}$ pertains to the interior. The right moving amplitude to the right of the barrier is

$$C_r = C_\ell \left[\left(1 - \frac{ik_0}{k}\right) e^{ia(k_0 - k)} + \frac{ik_0}{k} e^{iak_0} - e^{iak} \right]. \tag{94}$$

Consider now that the potential barrier has greater energy that the free particles. This gives

$$C_r = C_\ell \left[\left(1 - \frac{ik_0}{k}\right) e^{-a(k_0 e^{-k})} + \frac{ik_0}{k} e^{-ak_0} - e^{iak} \right]. \tag{95}$$

The modulus square of the quantum amplitude is

$$|C_r|^2 = calA^2 + |C_\ell|^2 \left[1 + \left(1 + 2\frac{k_0^2}{k^2}\right) e^{-2ak_0} + 2\mathcal{A}C_\ell e^{-ak_0} \right], \tag{96}$$

where $\mathcal{A} = A_2 + A_\ell(k_0/k - 1)$ With certain physical assumptions, such as C_r is zero if $a == 0$ The square of this amplitude is of the form

$$|C_r|^2 = 1 + e^{-2ak_0} - e^{-ak_0}, \tag{97}$$

with $|C_r|^2 = 0$ for $a = 0$ and as $a \to \infty |C_r|^2 = 1$. This latter condition is unphysical for the black hole beyond $a = 2m$, but for a physical barrier with this asymmetry between the k_r and k_ℓ this limit is of a classical impenetrable sphere. The physical approximation occurs for values of the step $a \simeq 1/k_0$.

This reflection coefficient models a thermal distribution of photons. If we interpret this as relativistic then $e^{-ak_0} \rightarrow e^{-Ea}$ for $E = pc$. We then then have the width playing the role of $a \rightarrow 1/kT$. We may then consider this sort of trapping region as something with an index of refraction that traps light. We may then consider the quantum state of interest as embedded in this medium. The quantum states in the interior of the black hole are contained in the inner timelike region, or on the inner event horizon.

The next step is to model a Reisner-Nordstrom black hole as an optical system. This will be done with the approximate metric

$$ds^2 = \left(1 - \frac{2m}{r} + \frac{Q^2}{r^2}\right)dt^2 - dr^2 - r^2 d\Omega^2, \tag{98}$$

where the $(1 - 2m/r + Q^2/r^2)^{-1}$ with the radial term is ignored. Dividing through by ds^2 gives

$$1 = \left(1 - \frac{2m}{r} + \frac{Q^2}{r^2}\right) - \left[\left(\frac{dr}{dt}\right)^2 - \left(\frac{d\Omega}{dt}\right)^2\right]\left(\frac{dt}{ds}\right)^2 \tag{99}$$

This a form of Lagrangian and the associated Hamiltonian is

$$H = \frac{1}{2}\left(\dot{r}^2 + r^2\dot{\phi}^2\right) - m/r + Q^2/r^2. \tag{100}$$

We are primarily interested in the radial evolution of a particle of mass μ falling inward. The quantum mechanical Hamiltonian of interest is

$$H_{op} = -\frac{\hbar^2}{2\mu}\partial_r^2 + 1 - m/r + Q^2/r^2. \tag{101}$$

for a wave function $\psi = e^{ikr - i\omega t}$. The unit in the potential is from the Schwarzschild metric and is retained to maintain the metric structure. For the stationary condition $H_{op}\psi = E\psi$ the wave vector is $k = \sqrt{2m/\hbar^2}\sqrt{E - V}$ for $V = -m/r + Q^2/r^2$. The condition for $r \in (m/2E - \sqrt{(m/2E)^2 - Q/E},\ m/2E + \sqrt{(m/2E)^2 - Q/E})$. Then just as with the potential step problem we have an interior region with real valued momentum.

This model of a black hole can be emulated in quantum dots. A quantum dot that holds a free quantum state in its interior, surrounded by a material with an effective potential can serve as a quantum processor or unit. It may then be possible to store quantum entangled states in this "black holes." If this is performed in a medium with light trapping properties, it is then possible to store entanglement information with time dilation on the access of this information. This may serve as a block or "on-off" gate to classical LOCC information in teleportation.

Materials that slow light or even stop it [38] may be fashioned into quantum dots. They could further potentially be fashioned into quantum cavities that would emulate a black hole, and further a black hole with timelike region in their timelike interior. This could in the future even serve as a type of memory storage system, and in a quantum mechanical setting are accessible only by those with the public key in LOCC.

8. Concluding remarks

This article is meant to illuminate the parallel between topics in the foundations of physics with other areas, such as computer science, material science, quantum optics, and additional physical properties and systems. As we find parallels between spacetime physics and quantum and optical systems we may emulate deep aspects of the universe. Conversely, we may develop new technologies that employ new platforms for communications and computing.

If computing systems involve analogues, discrete applications and other derivatives from physical foundations, this may extend the algorithmic complexity and power of computing technology. A black hole may be considered the ultimate processors of quantum states, and if emulated, however approximately, this may extend computing technology into new capabilities. The quantum dot may provide a model for a tiny quantum black hole and serve as circuit elements in different forms of quantum computer. With the structures that are presented here, the opportunity to explore a wide range of physical computing platforms is very large.

Author details

Lawrence B. Crowell
AIAS Budapest, Hu and Jefferson, Texas, United States of America

*Address all correspondence to: goldenfieldquaternions@gmail.com

References

[1] Lloyd S. The universe as quantum computer. Available from: https://arxiv.org/abs/1312.4455

[2] 'tHooft G. The cellular automaton interpretation of quantum mechanics. Fundamental Theories of Physics. 2016; **185** arxiv:1405.1548

[3] Hossenfelder S, Palmer T. Rethinking Superdeterminism. Frontiers in Physics. 2020;**8**:139 Available from: https://arxiv.org/abs/1912.06462

[4] Schneider P. p-adic Lie Groups. Berlin: Springer-Verlag; 2011

[5] Matiyasevich Y. Hilbert's 10th Problem. MIT Press Series in the Foundations of Computing. Foreword by Martin Davis and Hilary Putnam. Cambridge, MA: MIT Press; 1993

[6] Dvali G, Osmanov Z. Black holes as tools for quantum computing by advanced extraterrestrial civilizations. Available from: https://arxiv.org/abs/2301.09575

[7] Berry MV, Keating JP. H = xp and the Riemann zeros. In: Lerner IV, Keating JP, Khmelnitski DE, editors. Supersymmetry and Trace Formula: Chaos and Disorder. New York: Kluwer Academics/Plenum Publisher; 1999

[8] Edwards HM. Riemann Zeta Function. New York: Academic Press; 1974

[9] Susskind L. Three lectures on complexity and black holes. Available from: https://arxiv.org/pdf/1810.11563

[10] Claus P, Derix M, Kallosh R, Kumar J, Townsend PK, Van Proeyen A. Black holes and Superconformal mechanics. Physical Review Letters. 1998;**81**:4553-4556. Available from: https://arxiv.org/pdf/hep-th/9804177v3

[11] Galajinsky A. A variant of Schwarzian mechanics. Nuclear Physics B. 2018;**936**:661. Available from: https://arxiv.org/abs/1905.01935

[12] Siegel CL. Topics in Complex Function Theory. Vol. 2. Wiley-Interscience; 1971. Available from: ISBN 0-471-79080 X

[13] Berry MV, Keating JP. The Riemann zeros and eigenvalue asymptotics. SIAM Review. 1999;**41**:236

[14] Susskind L. Three lectures on complexity and black holes. Available from: https://static.ias.edu/pitp/2018/node/1796.html (arXiv:1810.11563v1)

[15] Liu P, Liu Z, Chen S, Ma X. Fundamental limitation on the detectability of entanglement. Physical Review Letters. 2022;**129**:503. Available from: https://arxiv.org/abs/2208.02518v2

[16] Danielson DL, Satishchandran G, Wald RM. Killing horizons decohere quantum superpositions. Available from: https://arxiv.org/abs/2301.00026v1

[17] Arora S, Barak B. Computational Complexity. Cambridge University Press; 2009. p. 230

[18] Ovsienko V, Tabachnikov S. What is the Schwarzian derivative? Notices of the AMS. 2009;**56**:1

[19] Newell A. Solitons in Mathematics and Physics. SIAM; 1985

[20] Wahlquist HD, Estabrook FB. Bäcklund transformations for solitons of

the Korteweg–de Vries equation. Physical Review Letters. 1973;**31**: 1386-1390

[21] Oppenheim J. Is it time to rethink quantum gravity?. Available from: https://arxiv.org/abs/2310.12221

[22] Lurie J. On the classification of topological field theories. AIM 2009-24. 2009. Available from: https://arxiv.org/abs/0905.0465v1

[23] van Raamsdonk M. Building up space-time with quantum entanglement. International Journal of Modern Physics D. 2009;**42**(14):2323-2329. Available from: https://arxiv.org/abs/1005.3035

[24] Maldacena J, Susskind L. Cool horizons for entangled black holes. Fortschritte der Physik. 2013;**61**(9): 781-811. Available from: https://arxiv.org/abs/1306.0533

[25] Crowell L, Corda C. Quantum hair on colliding black holes. Entropy. 2020; **22**(3):301. Available from: https://arxiv.org/abs/2007.01106v1

[26] Unruh WG, Schützhold R. On slow light as a black hole analogue. Physical Review D. 2003;**68** 2:024008. Available from: https://arxiv.org/abs/gr-qc/0303028

[27] Aaronson S, Watrous J. Closed Timelike curves make quantum and classical computing equivalent. Proceedings of the Royal Society A. 2008;**465**:631-647. Available from: https://arxiv.org/abs/0808.2669

[28] Bagger J, Wess J. Supersymmetry and Supergravity. Princeton University Press; 1983

[29] Efetov K. Supersymmetry in Disorder and Chaos. Cambridge University Press; 1997

[30] Kittel C. Introduction to Solid State Physics. New York: Wiley; 1996

[31] Dobardžić E, Dimitrijević M, Milovanović MV. Generalized Bloch theorem and topological characterization. Physical Review B. 2015;**91**:125424. Available from: https://arxiv.org/abs/1411.4920

[32] Hogarth ML. Does general relativity allow an observer to view an eternity in a finite time? Foundations of Physics Letters. 1992;**5**(2):173-181

[33] Soare R. Computability and recursion. Bulletin of Symbolic Logic. 1996;**2**(3):284-321

[34] Almheiri A, Engelhardt N, Marolf D, Maxfield H. The entropy of bulk quantum fields and the entanglement wedge of an evaporating black hole. JHEP. 2019;**2019**(12):63

[35] Khuri MA, Rainone JF. Black lenses in Kaluza-Klein matter. Physical Review Letters. 2023;**131**:041402

[36] Mulmuley KD. On P vs. NP, geometric complexity theory, and the Riemann hypothesis. Available from: https://arxiv.org/abs/0908.1936v2

[37] Coecke B. “The logic of entanglement”. Research report PRG-RR-03-12. 2003. Available from: arXiv:quant-ph/0402014

[38] Hau LV, Harris SE, Dutton Z, Behroozi CH. Light speed reduction to 17 meters per second in an ultracold atomic gas. Nature. 1999;**397**(6720):594-598

Section 2

Advanced Semiconductors

Chapter 4

Perspective Chapter: Advanced Semiconductors in Artificial Photosynthesis

Keran Jiao, Hansong Zhang, Xiaowu Gao, Ziwei Zhao, Pengfei Song and Yongjie Wang

Abstract

Artificial photosynthesis, which could utilize renewable solar energy to produce fuels, e.g., hydrogen, syngas, methanol, and so on, is regarded as a prospective alternative to the global energy and environment crisis. Semiconductor-based photocatalysts play a key role in such solar-to-fuel conversions. In this chapter, we will review the recent demonstrations of advanced semiconductors used in solar water splitting and carbon dioxide reduction. The requirements for the physical and chemical properties of photocatalytic semiconductors are discussed to distinguish the potential candidates. The demonstrations of such advanced semiconductors successfully driving solar water splitting with hydrogen production and carbon dioxide reduction with hydrocarbon fuels are summarized. Moreover, current bottleneck issues, possible strategies, and perspectives are proposed at the end. This chapter will be an important reference for the research in artificial photosynthesis using emerging semiconductor materials.

Keywords: semiconductors, photocatalysis, hydrogen, water splitting, carbon dioxide conversion, carbon neutrality, artificial photosynthesis

1. Introduction

Photosynthesis, which is a biological process, achieves superior conversion from light energy to reducing power (NADPH or NADH) and adenosine triphosphate (ATP). These chemicals are utilized to drive the fixation and reduction of carbon dioxide (CO_2) for the production of sugars [1]. The nature of photosynthesis processes in organisms with H_2O-reducing equivalents exemplifies the release of O_2, which contributes to virtually all production of O_2 in the atmosphere [2] (1). Building upon the promising aspects of artificial photosynthesis, this chapter provides a concise introduction to the principles and related components of semiconductor-based artificial photosynthesis. Furthermore, this chapter also categorizes and examines recent advancements in semiconductor technology for solar water splitting and carbon dioxide reduction. It discusses the necessary physical and chemical properties

IntechOpen

of photocatalytic semiconductors to identify promising candidates. Summaries of successful demonstrations utilizing these advanced semiconductors for solar water splitting and carbon dioxide reduction are provided. The chapter concludes by addressing current challenges, suggesting potential strategies, and offering perspectives on future developments.

$$CO_2 + 2H_2O \rightarrow (CH_2O) + H_2O + O_2 \quad (1)$$

In order to realize the sustainable development goals of the United Nations, artificial photosynthesis, which mimics the enzyme in natural biochemical photosynthesis, has been developed to convert CO_2 and H_2O (typically) into other chemical feedstocks [3, 4]. Overall, the vast majority of artificial photosynthesis research, especially for mimicking photosynthesis, relies on biomass-based catalytic methods, such as gasification and pyrolysis [5, 6]. While these systems are widely studied and have become important hydrogen production and CO_2 conversion alternatives, the detailed mechanisms and the cost are still challenging because the catalysts are often based on biochemical materials with different optimal conditions. Nanomaterials for artificial photosynthesis have witnessed tremendous development through structure designing based on detailed molecular orbitals and various morphologies emerging as leading active sites that can be manipulated and engineered toward the desired products [7]. A promising approach to improve the light-harvesting process features the application of advanced semiconductor materials for artificial photosynthesis [8]. Such semiconductors can be synthesized using established methods compatible with the semiconductor industry and offer an enhanced understanding of the light-harvesting mechanisms. This knowledge can serve as a basis for the band design in artificial photosynthesis and accelerate the passage of the semiconductor-based devices from the laboratory to applications.

Since the inceptive report by Fujishima [9], semiconductors with light-harvesting abilities in the ultraviolet–visible–near infrared (UV–vis–NIR) regions have attracted worldwide research interests to convert from classical fossil resources to renewable energy and alternative carbon sources [10]. In the studies of semiconductors, band theory is always used to describe the orbital-hybridized bands in potentials versus the vacuum level (E_{vac}). In this principle, the highest occupied orbital-hybridized band is defined as the valence band (VB); on the other hand, the lowest unoccupied is the conduction band (CB). The band gap describes the energy difference between VB and CB. For artificial photosynthesis, semiconductor-based candidates are attractive for their light response, charge dynamics, and surface chemistry, which makes them ideal active materials for governing the apparent solar-to-chemical energy conversions. The systematic analysis of these physical properties of semiconductor-based catalysts led to two main different counterparts for photocatalytic conversion. On this basis, suitable band gaps and carrier concentrations are needed to improve the range of the light absorption and the migration of electron-hole pairs, respectively, for the inter-band transition. Despite the importance of these physical properties in photocatalysis, surface photochemistry, such as bond activation and dissociation, also plays an important role in the conversion efficiency. In this way, tailor-made approaches were designed and used for artificial photosynthesis, enabling the improvement of the photocatalytic activity by means of structural modification and construction [11, 12].

Despite the great success of TiO_2-based materials in photocatalytic water splitting, a CdS/S^{2-} system was reported in 1979. The work by Inoue was based on the CdS photoanode, which is capable of CO_2 reduction [13], and therefore, a promising photochemical route by reducing CO_2 into various feedstocks has been proposed. Based on these studies, highly active semiconductor-based artificial photosynthesis has been developed with great success, accordingly. However, most semiconductor photocatalysts exhibited photoactivity only in the ultraviolet region, resulting in low light-harvesting efficiency and, consequently, low catalytic efficiency. Additionally, the timescales of the light-matter and chemical reactions are mismatched, exhibiting the loss of quantum efficiency that is difficult to control (femtoseconds-to-milliseconds), hindering the conversion efficiency [14]. The considerable diversity in semiconductor structures opens up promising avenues for strategies based on morphological and chemical design. Additionally, significant breakthroughs are expected in the upcoming years. In this chapter, we review methods for the solar water splitting and carbon dioxide reduction with advanced semiconductors. Mechanisms of semiconductor-based photocatalysts are categorized, and current bottleneck issues and design strategies toward these hindrances are presented. These promising directions can be the important references for switching to a green-developed artificial photosynthesis economy.

2. Principles and components of semiconductor-based artificial photosynthesis

For a thematic introduction, we will briefly introduce the basic principles of semiconductor-based photocatalysis and related components in these systems. We will also discuss the main common threads with classical photosynthesis, as well as the key differences. For natural photosynthesis, numerous different steps are available for artificial mimicking to drive the conversion of CO_2 and H_2O to carbohydrates and oxygen. In addition to a deeper understanding of the assemblies bioinspired in the molecular aspect, well-founded reports for chromophore for light harvesting can be accessed, making it comparatively easy to study the basic mechanisms [15, 16]. These molecules, such as porphyrin, are usually linked to the electron acceptors and secondary electron donors. In the excitation process, the photoinduced electron transfer generates a primary charge-separation state. The electron transferences segregate the redox equivalents in space, minimizing electronic coupling and slowing down the recombination of the charge-separated state. This property makes them suitable for applications in the photocatalytic processes [3].

While the basic mechanisms for natural photosynthesis were discussed as the representative explanation for photocatalysis, analogous approaches to mimicking natural photosynthesis are significantly more difficult since the reaction processes and charge separation processes usually show different rates, which are around two orders of magnitude. The huge differences can cause the recombination of the charged species and reduce the efficiency of the redox catalytic processes. Influenced by this aspect, this complexity of the reaction system is further increased by the different side reactions on the one hand and the other hand by the question, what influence does the byproduct have reactivity on the system. In response to the often-challenging mismatching rates between the catalytic reaction and charge creation/separation processes, more robust inorganic systems (semiconductor-based photocatalysis) have been developed to give a more stability and higher conversion rates compared to the organic and supramolecular systems. In this system, the reaction process of semiconductor photocatalysts can be illustrated into three steps (**Figure 1**):

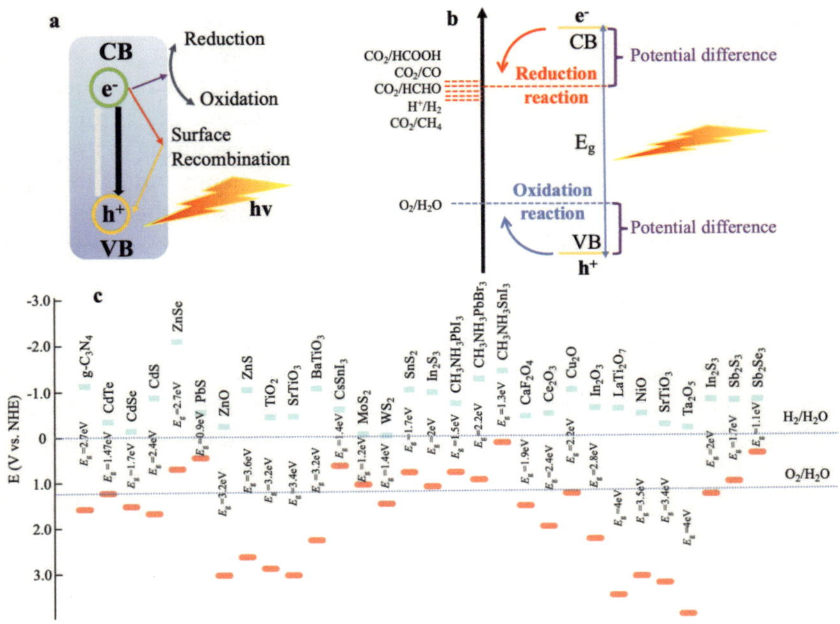

Figure 1.
(a) Schematic of the charge migration pathway in semiconductor-based artificial photosynthesis; (b) relationship between the redox potentials and the band structures in semiconductor photocatalysts [17]; (c) band edge positions of representative semiconductor materials [18–21].

(1) Under visible light irradiation, when the semiconductor photocatalyst absorbs light energy greater than its band gap, the electrons of the semiconductor are excited and transitions to the conduction band (CB), generating photogenerated electrons, while photogenerated holes are formed at the valence band (VB), creating photogenerated electron-hole pairs. (2) The migration of photogenerated electrons and holes to the surface of the photocatalyst happens. (3) Photogenerated electrons undergo the reaction with CO_2 or H^+ adsorbed on the semiconductor surface to drive the conversion of CO_2 and H_2O. In contrast, photogenerated holes undergo oxidation reactions with water or other electron sacrificial agents.

To develop new functional elements inspired by natural photosynthesis, such as the harvesting of sunlight photons, electron-hole separation with long time stabilities, and energy transduction. An artificial photosynthesis system should be composed of the following three main elements: (1) anode, (2) cathode, and (3) membrane, which can separate the two compartmental zones. In the semiconductor-based artificial photosynthesis system, the anode exposed to light irradiation can catalyze the oxidation of water. This structure can enhance the electron collections while remaining permeable to protons, facilitating the transportation of both electrons and protons to the cathode side through the electron transference and membranes. An important solution is the use of an electrocatalyst, which can promote the water electrolysis utilized by the semiconductor supplier. Due to the visible light activity and direct contact between the semiconductor and electrocatalyst, this method has been used by Swearer [22] to access various photocatalytic systems. The antenna effect enables the extension of the wavelength range of activity when the sensitizers for the semiconductor can withstand anodic oxidation conditions. Another important variant features 2D photonic materials, which are used for the absorption of visible light and the photon transfer process of energy [23]. However, a crucial factor for the anode in artificial photosynthesis lies in the need for efficient electron collection and rapid transportation of photogenerated protons. This process occurs through the anode itself and at the interface with the membrane. And for the cathode, the protons need to cross the interface to reach the catalytic sites for reaction. Therefore, the performance of the anode is significantly impacted by the effectiveness of transporting electrons and protons to the cathode, encompassing both external and internal electron conduction as well as a proton-permeable membrane. Consequently, the anode must possess a specific nanostructure and characteristics to ensure the efficiency of these processes and prevent limitations in the overall process rate. In addition to the anode, extensive research has been conducted on the cathode, solidifying its crucial role in the realm of artificial photosynthesis. A cathode, in its simplest form, comprises a conductive substrate which contacts with the membrane and is permeable to protons, including active centers facilitating the recombination of protons and electrons into H_2. In a more intricate approach, the cathode includes active sites capable of chemisorbing CO_2 and catalytically converting it, leveraging electrons and protons, into fuels or valuable chemicals [24]. Another important component is the membrane, which facilitates the rapid transport of protons (and potentially electrons, which are preferably transported externally either through a wire or physically separated elements within the membrane; the separation of electron and proton transference is essential to prevent recombination). Options for such a membrane include carbon nanotubes (CNTs) or TiO_2 connecting the electrodes, as well as ion-exchange resins, such as Nafion or other proton-conducting membranes. The membrane should also be customized to minimize proton (and possibly electron) transfer resistance, achieving a minimal thickness (ideally a few microns). Additionally, it should impede O_2 permeation to maintain H_2 evolution activity at the cathode. The membrane should maintain excellent contact with the anode and cathode, and efforts should be made to minimize resistance at the interface.

3. Water splitting for hydrogen production

In the following section, relevant strategies using catalytically active advanced semiconductor-based systems are categorized. Water splitting and reduction reactions with CO_2 were identified as major clusters in this field. Additionally, significant but rather scattered emerging directions will be outlined.

Artificial photosynthesis involving hydrogen, which is catalyzed by semiconductor-based catalysts, is of paramount importance in industry-scale chemical processes (e.g., the hydrogen production and decomposition of organic pollutants [25]), thanks to the excellent hydrogen economy, scalability, and recyclability of the semiconductor materials. Since the inceptive report by Baur and Rebmann [26], photo-assisted water splitting has been extensively studied with light irradiation. Since the identification of the Honda–Fujishima effect in 1972, numerous endeavors have been undertaken to find a photocatalyst in which both proton reduction and water oxidation reactions occur simultaneously at distinct sites [9, 16]. In this system, the semiconductor-based material can absorb photons whose energy is greater than or equal to the band gap, thus exciting electrons from VB into the CB and creating positively charged holes in the valence band. In the general water-splitting process, the Gibbs free energy (ΔG_0 = 237.13 kJ/mol) is positive, which makes the reaction cannot transpire spontaneously.

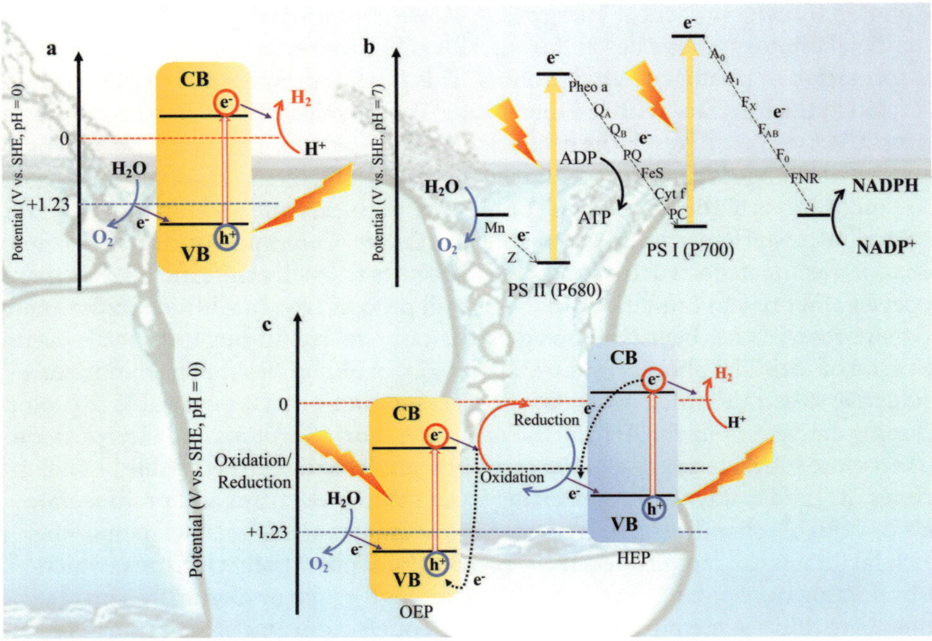

Figure 2.
(a) Single photocatalyst mechanisms in water-splitting system; (b) double excitation process in natural photosynthesis [16]; (c) double excitation system in semiconductor-based photosynthesis.

Photogenerated electrons and holes possess robust reduction and oxidation potentials, making them suitable for the following catalysis redox reactions on the surface (**Figure 2a**) [27]. Usually, water splitting necessitates two electrons or four holes to produce a H_2 or O_2, respectively. Consequently, the appropriate band gap sizes and the positions of the CB minimum and VB maximum are crucial for water-splitting process. In this system, the semiconductor-based materials need to have a CB minimum that features a more negatively positioned than the reduction potential of H^+/H_2. Additionally, the VB maximum should be more positively situated than the oxidation potential of O_2/H_2O (+1.23 V vs. standard hydrogen electrode (SHE), pH = 0). Thus, the minimum requirement for the energy of photon is 1.23 eV. Hence, the minimum theoretical photon energy requirement is 1.23 eV, corresponding to a wavelength of 1000 nm in the near-infrared range (1000 nm). Furthermore, the presence of overpotential becomes crucial to overcome the energy barrier in redox reactions involving water splitting. Therefore, the photocatalyst's band gap should exceed the requirement of 1.23 eV. Moreover, in the context of the light harvesting, around 52–55% of sunlight at Earth's surface consists of infrared light, 42–43% comprises visible light, and 3–5% is ultraviolet light [28]. Due to this, in order to obtain a reasonable efficiency in the artificial photosynthesis for water-splitting system, the band gap of semiconductors should be less than 3.1 eV to harness more visible light while maintaining a sufficiently large overpotential. However, certain sulfides and nitrides which feature the suitable band gap potentials for this system are often unstable because of the self-oxidation [29]. For instance, many (oxy) nitrides undergo self-oxidative deactivation to some extent, introducing the N 2p orbital in the VB formation, where photogenerated holes oxidize N^{3-} to N_2. This self-oxidative deactivation occurs competitively, and, in some cases, even with the use of appropriate cocatalysts, the water-splitting rate gradually decreases as the compound decomposes.

Unfortunately, regarding materials with narrower band gaps, like GaN-ZnO solid solutions, capable of decomposing water into H_2 and O_2 at 400–700 nm wavelength range, their efficiency remains unsatisfactory, which the quantum yield was only 0.14% in the 300–480 nm [30]. To date, there are only a few instances in the literature demonstrating monolithic water splitting in a single photocatalyst system in the 400–700 nm range [31, 32]. Due to the difficulty in designing both the size and positioning of the band gap, most reported visible-light-driven photocatalysts can only effectively perform the half-reactions of reduction or oxidation [33]. To obtain an efficient artificial photosynthesis system, a combination of oxygen evolution catalyst and hydrogen evolution catalyst method has been established in the field of water splitting by mimicking the two in natural photosynthesis (**Figure 2b**) and a double-excitation process in natural photosynthesis [34]. As shown in **Figure 2c**, with the connection of the electron mediator, the electrons excited in the CB of a hydrogen evolution photocatalyst engage in proton reduction, while the holes in the VB of an oxygen evolution photocatalyst participate in water oxidation, resulting in the evolution of H_2 and O_2, respectively. To complete the cycle and maintain charge neutrality, the CB electrons of the oxygen evolution photocatalyst traverse a relay/shuttle to unite with photo-generated holes in the VB of the hydrogen evolution photocatalyst. The two-stage photocatalytic system should have the conditions that: (1) an oxygen evolution photocatalyst capable of injecting electrons into hole acceptors, (2) a hydrogen evolution photocatalyst capable of receiving electrons from donors, and, importantly, (3) an electron relay system with appropriate potentials to suppress the following reactions. In such a dual excitation photocatalytic system, both the oxygen evolution catalyst and hydrogen evolution catalyst can be activated by visible light, as the band gap requirements for overall water-splitting processes are less demanding [16]. This theoretically allows for an efficient solar energy conversion (40%), in contrast to approximately 30% achievable with a single photocatalyst. Thanks to the reduced energy demand for each half-reaction compared to the higher energy requirement for driving the water-splitting process using a single photocatalyst. For example, oxide materials like WO_3, responsive to visible light, can facilitate oxygen evolution by reducing a mediator oxidant to the mediator reductant. Likewise, non-oxide semiconductors such as sulfides, nitrides, and dyes could contribute to hydrogen evolution if they can oxidize the mediator reductant to the mediator oxidant. Consequently, a broader range of semiconductors can find application in the two-stage system, even if they do not meet the stringent criteria for the overall water-splitting process as a single photocatalyst system. Due to this, the researchers usually focus on the dual photosynthesis system.

While the successful design of dual excitation for water-splitting systems was discussed as a promising design direction, the design approaches on artificial photosynthesis are significantly more difficult. The potential for O_2/H_2O is at +1.23 V. However, achieving gaseous oxygen production often requires a more positive valence band maximum position due to the overpotential in the chemical reaction [35]. The process of oxygen generation from water-splitting reaction is relatively slow because it necessitates 4 holes to form one oxygen molecule through water oxidation, whereas only two electrons are required for hydrogen production. Additionally, electrons transfer much more rapidly than holes in most oxide catalysts, which are predominantly n-type semiconductors, making the process more efficient. Therefore, a semiconductor with a deep VB maximum is essential to provide a substantial driving force. However, a drawback of this deep VB maximum, a common issue in many metal oxides, is the resulting large band gap, limiting the absorption of a significant portion of the visible spectrum from solar irradiation [36]. Only a handful of metal oxides, such as monoclinic WO_3 [37], $BiVO_4$ [38], and Bi_2MoO_6 [38], have been reported to efficiently carry out visible light range for half reactions like oxygen evolution. On the other hand, the advancement of the other H_2 evolution reaction is naturally indispensable for a dual two-excitation system.

Despite the abovementioned situation, in the young field of the artificial photosynthesis, the primary design directions for application were to gain the high overall efficiencies. To demonstrate the practical feasibility of water splitting, several approaches have been explored to enhance the efficiency of photocatalysts in visible light range. An important method is the use of cocatalyst loading, which can enhance the charge separation and facilitate the chemical reaction at the surface. In the case of most photocatalysts employed for traditional one-step water-splitting reactions, the co-catalysts that improve water reduction without catalyzing the formation of water from hydrogen and oxygen (such as NiO_x and Cr_2O_3/Rh) are commonly used to boost the efficiency of water splitting. Emerging design directions have also explored the simultaneous loading of an effective oxygen-evolving cocatalyst (e.g., Mn_3O_4 and IrO_2) with a hydrogen evolution co-catalyst [39]. The loading of cocatalysts significantly influences the efficiencies of the semiconductor-based photocatalytic systems. In such a dual photocatalyst system incorporating a redox reaction couple, the loading of co-catalysts that facilitate the reduction or oxidation of the mediator, rather than water, usually plays a dominant role in overall efficiency. For instance, the conversion of IO_3^- to I^- via 6 electrons frequently is the rate-determining step in the oxygen evolution side. Therefore, the loading of co-catalysts, such as PtO_x and RuO_2, can substantially improve the rate of oxygen evolution, thereby enhancing the overall efficiency of dual photocatalytic water splitting. Another important variant features heterojunction, which enables the adjustment of properties of the semiconductor materials for the improvement in the catalytic performance involving suppressed charge recombination. In contrast to the mixed powder of two semiconductor-based components, mechanically, the heterojunction structure demonstrated a twofold increase in activity, primarily attributed to a significantly more intimate contact at the interface. Besides the targeted modification and doping process, it should be noted that numerous morphologies will also influence the overall reaction performance. As an illustration, on the basis of the lanthanum and alkaline metal ions doping processes, the morphologies of $NaTaO_3$ can be finely adjusted from 4 to 0.1 μm, leading to a tenfold improvement in water-splitting activity. The meticulous control of morphology also contributes to the enhancement of dual excitation photocatalytic water-splitting systems. An intriguing instance is the "nature-inspired" architecture of the $CdS/Au/TiO_2$ system [40]. Ding and co-workers [41] explored the wing scale of a black butterfly and recognized that a quasi-honeycomb morphology could significantly enhance light absorbance. By synthesizing TiO_2 with a QH structure using the black butterfly's wing as an organic template, the $CdS/Au/TiO_2$ system showed a 2.3 times increase compared to the plate architecture-based TiO_2, underscoring the significance of morphology control.

4. Photocatalytic CO_2 reduction reaction

While the works mentioned above represent the benchmark light-harvesting mechanisms for artificial photosynthesis systems, the studies are mostly limited to water splitting [42, 43]. An impressive array conversion, for instance, CO_2 reduction, could be improved by such systems, thus testifying the adaptability of the photocatalytic semiconductor-based devices when employed in different photosynthesis routes. In the following section, relevant design directions in CO_2 conversions are introduced and were identified as major design directions in the field.

As above mention, the cathode is the active center for CO_2 adsorption and reduction. The primary challenge lies in achieving selective conversion, given the closely similar reduction potentials of CO_2 to various product types. Consequently, attaining high Faradaic selectivity and controlling the selectivity in the photocatalytic reduction of CO_2 to the desired product poses significant challenges. It is also evident that, to prevent the formation of high-energy intermediates, a simultaneous transfer of multiple electrons and protons is required—a characteristic shared with other catalytic reactions [44]. The one-electron reduction of carbon dioxide to the $\bullet CO_2$ anion radical is a straightforward reaction. However, the high potential required to produce this product suggests the necessity of achieving a one-step (direct) electron/proton addition, such as a multi-electron transfer, to form an intermediate leading to the desired product (**Table 1**). For instance, the carbon dioxide reduction to methanol involves a 6-electron transfer, while isopropanol formation requires an 18-electron transfer. Despite its difficulty and low productivity, metal nanoparticle-supported CNTs have demonstrated the possibility of generating C_x products through CO_2 electroreduction, including C_3H_8O [47].

Therefore, to achieve the efficient reduction of CO_2, the preparation and modification of semiconductor photocatalysts are crucial. The photocatalytic reduction of CO_2 is influenced by the band gap structure of the semiconductor catalyst itself. The catalyst band gap is quite important, as per the band-energy photon absorption wavelength formula $E_g = 1240/\lambda$. It is understood that the larger the absorption edge of the semiconductor photocatalyst, the narrower the corresponding band gap width. Thus, when the solar energy absorbed by the semiconductor exceeds its band gap width can the photocatalyst generate and separate electron-hole pairs. For instance, this phenomenon was discussed by Zhao and others under the term band gap modification [48]. They decompose upon the N-doping process, in which the N atoms enable the substitution of O in TiO_2 lattices. The success of the strategy can achieve the redshift of the light harvesting and the reduction of the band gap, which reflected in a 4 times higher activity in photocatalytic CO_2 reduction reaction, compared to pure TiO_2 nanotubes. Additionally, not all the photo-generated electron-hole pairs

Half-reactions of CO_2 reduction	Potentials (V vs. SHE, pH = 7)
$CO_2(g) + e^- \rightarrow CO_2^{\bullet -}$	$E^0 = -1.90V$
$CO_2(g) + 2e^- + 2H^+ \rightarrow HCOOH(l)$	$E^0 = -0.61V$
$CO_2(g) + 2e^- + 2H^+ \rightarrow CO(g) + H_2O(l)$	$E^0 = -0.53V$
$CO_2(g) + 4e^- + 4H^+ \rightarrow HCHO(l) + H_2O(l)$	$E^0 = -0.48V$
$CO_2(g) - 6e^- + 6H^+ \rightarrow CH_3OH(l) + H_2O(l)$	$E^0 = -0.38V$
$CO_2(g) + 8e^- + 8H^+ \rightarrow CH_4(g) + 2\,H_2O(l)$	$E^0 = -0.24V$
$2CO_2(g) + 12e^- + 12H^+ \rightarrow CH_3CH_2OH(l) + 3H_2O(l)$	$E^0 = 0.08V$
$2CO_2(g) + 12e^- + 12H^+ \rightarrow C_2H_4(g) + 4H_2O(l)$	$E^0 = 0.06V$

Table 1.
Potentials for half-reactions of CO_2 reduction [3, 45, 46].

produced in the above process can undergo photocatalytic reactions. Photocatalytic reactions require that the CB and VB of the semiconductor photocatalyst satisfy the reduction potential of CO_2 and the oxidation potential of H_2O, respectively. The more negative the position of the conduction band, the stronger the reduction capability; conversely, the more positive the valence band position, the stronger the oxidation capability (**Figure 1c**) [49]. In response to these two important properties in the process of preparing semiconductor photocatalysts for CO_2 reduction reaction, one should not only consider a narrower band gap or the semiconductor's strong redox ability but also take into account the combined influence of both factors. For example, the engineered tungsten oxide-based system is one of the most attractive methods for the CO_2 conversion under photocatalytic technology. Despite the visible light response and suitable band structure are promising for artificial photosynthesis, especially for CO_2 reduction [50]. The multiple forms such as WO_3, $W_{18}O_{49}$ (or $WO_{2.72}$), and $WO_3{\cdot}0.33H_2O$ and engineerable of the surfaces and interfaces could be effectively utilized to develop a negative conduction band position. Based on this, the use of Co atom for surface modification of $W_{18}O_{49}$ nanowires was reported by Zhang. They showed that the modified single Co atom can not only act as the active centers for CO_2 reduction but develop a more negative flat band potential applied in the redox system [51]. Similarly, a set of metal-free systems with the visible light response and strongly negative conduction band potential, such as graphitic C_3N_4 and 1,2,3-trimethylbenzimidazoline (BIH) were evaluated and exhibited promising efficiency and selectivity in photocatalytic CO_2 reduction [52–54]. Based on adjustment experiments in combination with band gap and band position, the visible response photocatalyst with excellent reduction capability can be postulated.

In addition, improving the recombination/decomposition rate of photogenerated electron–hole pairs within the semiconductor is still the crucial factor for optimizing photocatalytic performance. The recombination of photogenerated electron–hole pairs generally occurs through bulk recombination inside the material and surface recombination at the material's surface. During the photocatalytic process, photogenerated electrons and holes migrate to the CB and VB of the semiconductor, respectively. However, due to Coulombic attraction, recombination is prone to occur, acting as an impediment to their participation in the photocatalytic reduction of CO_2. In order to enhance the reduction activity, semiconductor heterojunctions can be helpful, which often feature the large-scale electron–hole pair separation and efficient charge migration [55]. Creating a composite structure with two or more semiconductors at the atomic level results in a band-bending effect at the interface due to differences in band edges or Fermi levels between the components. This band bending at the atomic-level interface facilitates the transfer of electrons and holes between the components in opposite directions. The separation of charge carriers across the interface significantly diminishes their recombination rate. On the other hand, the separation of reaction active sites in spatial also promotes forward reactions and inhibits backward ones, similar to the dual excitation processes in water splitting. This suggests a wide application prospect for CO_2 reduction [46]. For example, the TiO_2/CdS photocatalyst developed by Spanhel et al. in 1987 stands out as a quintessential example of a double-charge transfer heterogeneous photocatalyst. The observed blue shift in the absorption spectrum of ZnO indicates the injection of photogenerated electrons from the CB of CdS into the CB of the attached TiO_2 under visible light irradiation [56]. However, in general type II heterojunctions shown in **Figure 3c**, it may be necessary to sacrifice the more negative CB and more positive VB, resulting in a catalytic composite that cannot effectively reduce CO_2 into high reduction potential products, thereby compromising the catalytic performance

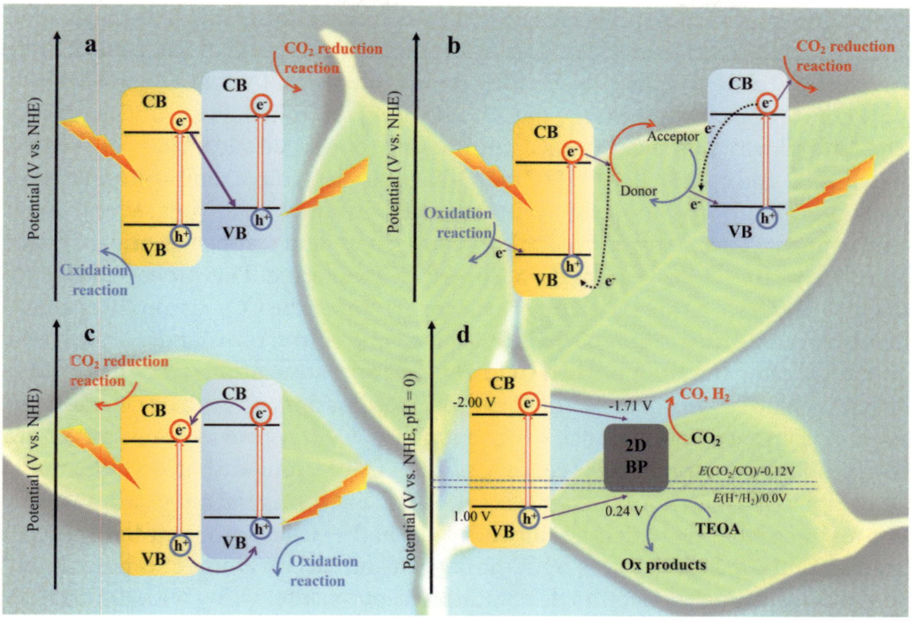

Figure 3.
(a) & (b) photocatalytic CO_2 reduction process in Z-scheme system; (c) photocatalytic CO_2 reduction process in type II heterojunction; (d) photocatalytic CO_2 reduction process in type I heterojunction with the sacrificial agent addition [57].

of the photocatalyst [58]. Similarly, a set of two processes excitation system (Z-scheme mode, **Figure 3a,b**) with one reduction semiconductor and one oxidation semiconductor has been developed and enabled simultaneously facilitate light response and maintain the strong reducibility and oxidizability of the reduction and oxidation semiconductors, respectively. Additionally, due to the Z-scheme system, the significantly enhanced charge transfer rate effectively inhibits the recombination of photoinduced electron–hole pairs, thus enhancing the efficiency of separating effective charge carriers. Furthermore, the separation of charge carriers can further promote the spatial segregation of active sites for reduction and oxidation, ensuring specific photocatalytic reactions occur at their respective active sites [46, 58]. For example, Bai et al. developed a Z-scheme g-C_3N_4/$Bi_4O_5I_2$ photocatalytic system for CO_2 reduction. The optimized g-C_3N_4/$Bi_4O_5I_2$ system exhibited an excellent CO which is 7.9 and 2.3 times higher than those achieved using pure g-C_3N_4 and $Bi_4O_5I_2$, respectively, in the absence of a cocatalyst and sacrificial reagent [59]. Another important method for the adjustment of recombination of charge carriers involves the sacrificial agent. For instance, Zhou et al. prepared a two-dimensional black phosphorus (2D BP) composite of 2D BP/g-C_3N_4 at extremely low temperatures (**Figure 3d**). The prepared g-C_3N_4 has CB and VB energies of −2.0 and 1.0 eV, respectively, which are lower and higher than BP's CB (−1.71 eV) and VB (0.24 eV), forming a typical type I heterojunction, which both electrons and holes are transferred to the same semiconductor, and this is unfavorable for the spatial separation of electrons and holes, leading to a higher probability of electron-hole recombination. In their studies, the adsorption of 2D BP is conducive to the separation of photogenerated holes and electrons, and the timely addition of the hole sacrificial agent triethanolamine (TEOA) effectively consumes the photogenerated holes, greatly suppressing the recombination of charge carriers [57].

Additionally, as shown in **Figure 4a**, surface defects on the catalyst are considered the most active catalytic reaction sites, altering the electron structure of the catalyst surface, enhancing local charge density, facilitating strong interactions with CO_2 molecules, reducing the chemical bond energy of CO_2, adjusting the light harvest, and accelerating the catalytic reaction. For instance, the presence of oxygen vacancies (VOs) in the hybrid of ZnO and carbon dots (OD-ZnO/C-N_2–600) serves as a multi-purpose. It not only extends the adsorption spectrum into the near-infrared region but also enhances CO_2 enrichment. These oxygen vacancies, hosting abundant trapped electrons, facilitate the injection of free electrons into the adsorbed CO_2 molecule. This process induces the bending of the O–C–O angle, leading to the formation of the CO_2^- anion, the extent of which also plays an important role in influencing the electron transfer dynamics between the semiconductor and the adsorbed CO_2. The potential adsorption structures of CO_2 primarily include oxygen coordination, carbon coordination, and mixed coordination. In terms of oxygen coordination, each oxygen atom in CO_2 may provide lone pair electrons to electron-deficient atoms on the catalyst surface. Concerning carbon coordination, the carbon atom in CO_2 can also act as an electron-deficient center, forming carbonate-like structures by combining with lone pair electrons on the surface atoms of the catalyst. In contrast, for mixed coordination, both the oxygen and carbon atoms in carbon dioxide can simultaneously act as electron donors and acceptors, interacting with the catalyst surface. Consequently, photocatalysts with oxygen vacancies, hydroxyl groups [60]

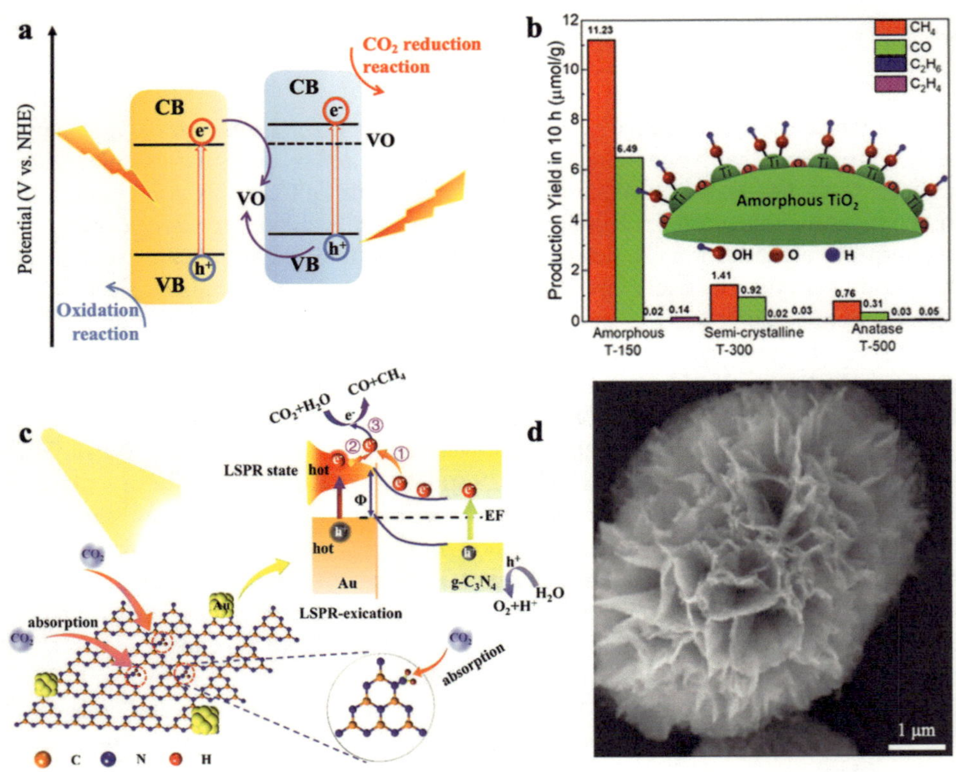

Figure 4.
(a) Defective heterojunction for CO_2 photoreduction; (b) hydroxyl groups on TiO_2. Copyright (2020), with permission from ACS publications [60]; (c) amine-modified g-C_3N_4 decorated with Au nanoparticles. Copyright (2020), with permission from Elsevier [61]; (d) ball-flower-like NiO/g-C_3N_4 heterojunction. Copyright (2018), with permission from Elsevier [62].

(**Figure 4b**), or amino groups [61] (**Figure 4c**) are more prone to CO_2 adsorption. Therefore, the adsorption and activation characteristics of CO_2 molecules on the catalyst surface determine the activity and product selectivity of the catalytic reaction. Similar to water-splitting reaction, by altering the morphology of the catalyst to increase the surface area and active sites for catalytic reactions, it is also possible to enhance the ability of CO_2 capture and adsorption, reduce the resistance of electron transfer, and thereby improve the photocatalytic performance and efficiency. Tang et al. successfully prepared a ball-flower-like NiO/g-C_3N_4 heterojunction structure, which is shown in **Figure 4d** through a hydrothermal doping and calcination process. Due to the high surface area and efficient separation of photogenerated charge carriers in this structure, the photocatalytic activity of the composite catalyst for CO_2 is 2.5 times that of pure g-C_3N_4 and 7.6 times that of pure NiO [62].

5. Conclusion and outlook

Fifty years after the successful establishment of the effect by Fujishima, artificial photosynthesis has received tremendous attention in chemistry. Developing an "artificial photosynthesis" system that mimics the energy collection process of a natural leaf holds the potential to address sustainability issues in energy. To overcome the intermittent nature of solar energy, it is crucial to devise systems capable of directly capturing CO_2 and converting it into liquid solar fuels for convenient storage. While H_2 production serves as a necessary intermediate step, challenges in gas storage underscore the importance of focusing on liquid solar fuels. However, for artificial photosynthesis to surpass their natural counterparts, they should exhibit excellent solar-to-chemical conversion efficiency, directly yield compounds and fuels for industrial, and be robust and easily constructible.

In this chapter, we delve into recent advancements in the utilization of sophisticated semiconductors for solar water splitting and carbon dioxide reduction. We also explore the essential physical and chemical properties required for photocatalytic semiconductors to distinguish potential candidates. Summarizing the successful demonstrations of these advanced semiconductors in driving solar water splitting for hydrogen production and carbon dioxide reduction to hydrocarbon fuels also became a focal point. Additionally, we have addressed current bottleneck issues and propose potential strategies.

In terms of potential applications, there is still a considerable journey ahead to enhance efficiency, reduce costs, and make large-scale viability a reality. Recent advancements have been discussed to showcase intriguing concepts in system integration, primarily. However, the hybrid device needs to demonstrate energy efficiency and productivity advantages compared to the conventional devices, which are already commercialized. Despite significant interest, both from the scientific community and the general public, much more research effort is essential. The development of semiconductor-based artificial photosynthesis stands as a significant human endeavor, akin to the moon landing, requiring the assembly of multidisciplinary competencies in large collaborative projects.

Acknowledgements

This research was funded by the National Natural Science Foundation of China (52102162), the Guangdong Basic and Applied Basic Research Foundation

(2022A1515011794), and the Shenzhen Constantly-Supported Project for Universities and Colleges (GXWD20231130110722002).

Conflict of interest

The authors declare no competing financial interest.

Author details

Keran Jiao[1,2], Hansong Zhang[1], Xiaowu Gao[1,3], Ziwei Zhao[1,3], Pengfei Song[2] and Yongjie Wang[1,3]*

1 Guangdong Provincial Key Laboratory of Semiconductor Optoelectronic Materials and Intelligent Photonic Systems, Shenzhen Key Laboratory of Advanced Functional Carbon Materials Research and Comprehensive Application Harbin Institute of Technology, Shenzhen, China

2 School of Advanced Technology, Xi'an Jiaotong-Liverpool University, Suzhou, China

3 National Key Laboratory of Science and Technology on Advanced Composites in Special Environments, Harbin Institute of Technology, Harbin, China

*Address all correspondence to: yjwang@hit.edu.cn

References

[1] Dey PM, Harborne JB. Plant Biochemistry. London: Academic Press; 1997

[2] Johnson MP. Photosynthesis. Essays in Biochemistry. 2016;**60**:255-273

[3] Perathoner S, Centi G. Artificial leaves using sunlight to produce fuels. In: Studies in Surface Science and Catalysis. Netherlands: Elsevier; 2020. pp. 415-430

[4] Lu Y, Nakicenovic N, Visbeck M, et al. Policy: Five priorities for the UN sustainable development goals. Nature. 2015;**520**:432-433

[5] Tanksale A, Beltramini JN, Lu GM. A review of catalytic hydrogen production processes from biomass. Renewable and Sustainable Energy Reviews. 2010;**14**:166-182

[6] Alstrum-Acevedo JH, Brennaman MK, Meyer TJ. Chemical approaches to artificial photosynthesis. 2. Inorganic Chemistry. 2005;**44**:6802-6827

[7] Kim D, Kley CS, Li Y, et al. Copper nanoparticle ensembles for selective electroreduction of CO_2 to C_2–C_3 products. Proceedings of the National Academy of Sciences of the United States of America. 2017;**114**:10560-10565

[8] Zhang N, Xiong Y. Plasmonic semiconductors for advanced artificial photosynthesis. Advanced Sensor and Energy Materials. 2023;**2**:100047

[9] Fujishima A, Honda K. Electrochemical photolysis of water at a semiconductor electrode. Nature. 1972;**238**:37-38

[10] Centi G, Perathoner S. Chemistry and energy beyond fossil fuels. A perspective view on the role of syngas from waste sources. Catalysis Today. 2020;**342**:4-12

[11] Cestellos-Blanco S, Zhang H, Kim JM, et al. Photosynthetic semiconductor biohybrids for solar-driven biocatalysis. Nature Catalysis. 2020;**3**:245-255

[12] Wang H, Zhang L, Chen Z, et al. Semiconductor heterojunction photocatalysts: Design, construction, and photocatalytic performances. Chemical Society Reviews. 2014;**43**:5234-5244

[13] Inoue T, Watanabe T, Fujishima A, et al. Investigation of CdS photoanode reaction in the electrolyte solution containing sulfide ion. Bulletin of the Chemical Society of Japan. 1979;**52**:1243-1250

[14] Corby S, Rao RR, Steier L, et al. The kinetics of metal oxide photoanodes from charge generation to catalysis. Nature Reviews Materials. 2021;**6**:1136-1155

[15] Barber J, Tran PD. From natural to artificial photosynthesis. Journal of the Royal Society Interface. 2013;**10**:20120984

[16] Wang Y, Suzuki H, Xie J, et al. Mimicking natural photosynthesis: Solar to renewable H_2 fuel synthesis by Z-scheme water splitting systems. Chemical Reviews. 2018;**118**:5201-5241

[17] Huo H, Wu F, Kan E, et al. Overall photocatalytic CO_2 reduction over heterogeneous semiconductor photocatalysts. Chemistry – A European Journal. 2023;**29**:e202300658

[18] Zhou H, Qu Y, Zeid T, et al. Towards highly efficient photocatalysts using

semiconductor nanoarchitectures. Energy & Environmental Science. 2012;**5**:6732-6743

[19] Li J, Wu N. Semiconductor-based photocatalysts and photoelectrochemical cells for solar fuel generation: A review. Catalysis Science & Technology. 2015;**5**:1360-1384

[20] Voiry D, Shin HS, Loh KP, et al. Low-dimensional catalysts for hydrogen evolution and CO_2 reduction. Nature Reviews Chemistry. 2018;**2**:0105

[21] Lu Q, Yu Y, Ma Q, et al. 2D transition-metal-dichalcogenide-nanosheet-based composites for photocatalytic and electrocatalytic hydrogen evolution reactions. Advanced Materials. 2016;**28**:1917-1933

[22] Swearer DF, Zhao H, Zhou L, et al. Heterometallic antenna–reactor complexes for photocatalysis. Proceedings of the National Academy of Sciences. 2016;**113**:8916-8920

[23] Centi G, Passalacqua R, Perathoner S, et al. Oxide thin films based on ordered arrays of 1D nanostructure. A possible approach toward bridging material gap in catalysis. Physical Chemistry Chemical Physics. 2007;**9**:4930-4938

[24] Kondratenko EV, Mul G, Baltrusaitis J, et al. Status and perspectives of CO_2 conversion into fuels and chemicals by catalytic, photocatalytic and electrocatalytic processes. Energy & Environmental Science. 2013;**6**:3112-3135

[25] Ismael M. Enhanced photocatalytic hydrogen production and degradation of organic pollutants from Fe(III) doped TiO_2 nanoparticles. Journal of Environmental Chemical Engineering. 2020;**8**:103676

[26] Baur E, Rebmann A. Über Versuche zur Photolyse des Wassers. Helvetica Chimica Acta. 1921;**4**:256-262

[27] Kudo A, Miseki Y. Heterogeneous photocatalyst materials for water splitting. Chemical Society Reviews. 2009;**38**:253-278

[28] Chen C, Ma W, Zhao J. Semiconductor-mediated photodegradation of pollutants under visible-light irradiation. Chemical Society Reviews. 2010;**39**:4206-4219

[29] Chen D, Ye J. Photocatalytic H_2 evolution under visible light irradiation on $AgIn_5S_8$ photocatalyst. Journal of Physics and Chemistry of Solids. 2007;**68**:2317-2320

[30] Maeda K, Takata T, Hara M, et al. GaN:ZnO solid solution as a photocatalyst for visible-light-driven overall water splitting. Journal of American Chemical Society. 2005;**127**:8286-8287

[31] Hu C, Chen F, Wang Y, et al. Exceptional cocatalyst-free photo-enhanced piezocatalytic hydrogen evolution of carbon nitride nanosheets from strong in-plane polarization. Advanced Materials. 2021;**33**:2101751

[32] Ma X, Cheng H. Synergy of nitrogen vacancies and intercalation of carbon species for enhancing sunlight photocatalytic hydrogen production of carbon nitride. Applied Catalysis B: Environmental. 2022;**314**:121497

[33] Konta R, Ishii T, Kato H, et al. Photocatalytic activities of noble metal ion doped $SrTiO_3$ under visible light irradiation. The Journal of Physical Chemistry B. 2004;**108**:8992-8995

[34] Ferreira KN, Iverson TM, Maghlaoui K, et al. Architecture of the

photosynthetic oxygen-evolving center. Science. 2004;**303**:1831-1838

[35] Turner JA. A realizable renewable energy future. Science. 1999;**285**:687-689

[36] Moniz SJA, Shevlin SA, Martin DJ, et al. Visible-light driven heterojunction photocatalysts for water splitting – A critical review. Energy & Environmental Science. 2015;**8**:731-759

[37] Sayama K, Mukasa K, Abe R, et al. A new photocatalytic water splitting system under visible light irradiation mimicking a Z-scheme mechanism in photosynthesis. Journal of Photochemistry and Photobiology A: Chemistry. 2002;**148**:71-77

[38] Kato H, Hori M, Konta R, et al. Construction of Z-scheme type heterogeneous photocatalysis systems for water splitting into H_2 and O_2 under visible light irradiation. Chemistry Letters. 2004;**33**:1348-1349

[39] Xiong A, Yoshinaga T, Ikeda T, et al. Effect of hydrogen and oxygen evolution cocatalysts on photocatalytic activity of GaN:ZnO. European Journal of Inorganic Chemistry. 2014;**2014**:767-772

[40] Kudo A, Niishiro R, Iwase A, et al. Effects of doping of metal cations on morphology, activity, and visible light response of photocatalysts. Chemical Physics. 2007;**339**:104-110

[41] Ding L, Zhou H, Lou S, et al. Butterfly wing architecture assisted CdS/Au/TiO_2 Z-scheme type photocatalytic water splitting. International Journal of Hydrogen Energy. 2013;**38**:8244-8253

[42] Bensaid S, Centi G, Garrone E, et al. Towards artificial leaves for solar hydrogen and fuels from carbon dioxide. ChemSusChem. 2012;**5**:500-521

[43] Centi G, Perathoner S. Chemical Energy Storage. In: Schlögl R, editor. Germany: De Gruyter; 2013. pp. 379-400

[44] Perathoner S, Centi G. Catalysis for solar-driven chemistry: The role of electrocatalysis. Catalysis Today. 2019;**330**:157-170

[45] Ross MB, De Luna P, Li Y, et al. Designing materials for electrochemical carbon dioxide recycling. Nature Catalysis. 2019;**2**:648-658

[46] Zhang W, Mohamed AR, Ong W-J. Z-scheme photocatalytic systems for carbon dioxide reduction: Where are we now? Angewandte Chemie International Edition. 2020;**59**:22894-22915

[47] Ampelli C, Centi G, Passalacqua R, et al. Synthesis of solar fuels by a novel photoelectrocatalytic approach. Energy & Environmental Science. 2010;**3**:292-301

[48] Zhao Z, Fan J, Wang J, et al. Effect of heating temperature on photocatalytic reduction of CO_2 by N–TiO_2 nanotube catalyst. Catalysis Communications. 2012;**21**:32-37

[49] Yan Y, Gu J, Zeitler EL, et al. Photoelectrocatalytic reduction of carbon dioxide. In: Carbon Dioxide Utilisation. Netherlands: Elsevier; 2015. pp. 211-233

[50] Yang G, Zhu X, Cheng G, et al. Engineered tungsten oxide-based photocatalysts for CO_2 reduction: Categories and roles. Journal of Material Chemistry A. 2021;**9**:22781-22809

[51] Zhang H, Wang Y, Zuo S, et al. Isolated cobalt centers on $W_{18}O_{49}$ nanowires perform as a reaction switch for efficient CO_2 photoreduction.

Journal of American Chemistry Society. 2021;**143**:2173-2177

[52] Xie W, Xu J, Md Idros U, et al. Metal-free reduction of CO_2 to formate using a photochemical organohydride-catalyst recycling strategy. Nature Chemistry. 2023;**15**:794-802

[53] Fu J, Zhu B, Jiang C, et al. Hierarchical porous O-doped g-C_3N_4 with enhanced photocatalytic CO_2 reduction activity. Small. 2017;**13**:1603938

[54] Ye S, Wang R, Wu M-Z, et al. A review on g-C_3N_4 for photocatalytic water splitting and CO_2 reduction. Applied Surface Science. 2015;**358**:15-27

[55] Tong Z, Yang D, Sun Y, et al. Tubular g-C_3N_4 isotype heterojunction: Enhanced visible-light photocatalytic activity through cooperative manipulation of oriented electron and hole transfer. Small. 2016;**12**:4093-4101

[56] Spanhel L, Weller H, Henglein A. Photochemistry of semiconductor colloids. 22. Electron ejection from illuminated cadmium sulfide into attached titanium and zinc oxide particles. Journal of the American Chemical Society. 1987;**109**:6632-6635

[57] Zhou G, Yang J, Zhu X, et al. Cryo-induced closely bonded heterostructure for effective CO_2 conversion: The case of ultrathin BP nanosheets/g-C_3N_4. Journal of Energy Chemistry. 2020;**49**:89-95

[58] Huang D, Chen S, Zeng G, et al. Artificial Z-scheme photocatalytic system: What have been done and where to go? Coordination Chemistry Reviews. 2019;**385**:44-80

[59] Bai Y, Ye L, Wang L, et al. G-C_3N_4/$Bi_4O_5I_2$ heterojunction with I_3^-/I^- redox mediator for enhanced photocatalytic CO_2 conversion. Applied Catalysis B: Environmental. 2016;**194**:98-104

[60] Kharade AK, Chang S. Contributions of abundant hydroxyl groups to extraordinarily high photocatalytic activity of amorphous titania for CO_2 reduction. The Journal of Physical Chemistry C. 2020;**124**:10981-10992

[61] Li F, Zhou H, Fan J, et al. Amine-functionalized graphitic carbon nitride decorated with small-sized Au nanoparticles for photocatalytic CO_2 reduction. Journal of Colloid and Interface Science. 2020;**570**:11-19

[62] Tang J, Guo R, Zhou W, et al. Ball-flower like NiO/g-C_3N_4 heterojunction for efficient visible light photocatalytic CO_2 reduction. Applied Catalysis B: Environmental. 2018;**237**:802-810

Chapter 5

Effects of Localization Length and Spatial Disorder on a Charge Carrier Mobility in Organic Disordered Semiconductors

Seyfan Kelil Shukri and Lemi Demeyu

Abstract

We investigate the transport properties of charge carriers in disordered organic semiconductors, with a specific focus on analyzing and determining charge carrier mobility. Our approach is based on the understanding that charge transport occurs through incoherent hopping of charge carriers across localized states. To model this behavior, we utilize a mobility expression that relates charge carrier hopping to thermal activation, specifically employing the Miller-Abrahams expression to describe the hopping rate. To generate data and analyze charge carrier mobility, we employ kinetic Monte Carlo simulation methods. By varying the applied electric field, temperature, localization length, and parameters related to spatial and energetic disorder, we obtain insights into the mobility behavior of charge carriers. Notably, we consider the impact of spatial disorder, which allows us to distinguish the effects of electric field and density, factors that are inseparable in experimental measurements. We compare our findings with experimental results and also investigate the influence of density and electric field on mobility at different temperatures, both in the presence and absence of spatial disorder. Our analysis highlights the significance of the spatial disorder parameter and the localization length in understanding the effects of the electric field on charge carrier mobility. Furthermore, we assess the commonly used value of the localization length, typically considered as $0.1b$, where b represents the lattice parameter.

Keywords: charge transport, charge carrier-dependent mobility, localization length, temperature-dependent mobility, spatial disorder

1. Introduction

Organic polymers that exhibit semiconducting properties due to weak π bonds along the polymer chain have received widespread attention in the scientific community because they are easy to fabricate and use in electrical devices, such as: organic light-emitting diodes (OLEDs) [1], organic field-effect transistors (OFETs) [2–4], and organic solar cells [5]. The performance of organic disordered semiconductors (ODSs) is based on the transport properties of charge carriers in the semiconductors that can be used for electronic devices. The main transport parameter is the drift mobility μ of charged carrier, which is much smaller than those of its inorganic counterparts. Therefore, to better utilize ODS as an active material in electronic devices, the transport properties of charge carriers have been widely studied [6–12], especially the dependence of mobility on temperature (T), electric field intensity (F), and density of charged particles (p) have been extensively studied in both experimental methods [13–20] and theoretically [6, 9, 21–25]. However, there is no consensus on an expression that uniformly describes the characteristics of the charge carrier mobility in various disordered organic semiconducting polymers.

Charge transport in ODS is due to the tunneling (or hopping) of charge carriers between localized states, which are considered to consist of conjugated polymer chain segments that are assumed to be randomly distributed in space. A hopping takes place from one state at an appointed energy level to another nearby state with the assistance of a phonon. Phonons are absorbed or emitted, so energy is conserved during charge carrier hopping. In ODS, the energies of charge carrier on localized states are assumed to have a Gaussian distribution [6]. The density of states (DOS) of this distribution is approximately given by

$$\rho(\varepsilon) = \frac{N}{\sqrt{2\pi\sigma^2}} \exp\left(-\frac{\varepsilon^2}{2\sigma^2}\right), \tag{1}$$

where N is the total concentrations; ε is the energy of a charge carrier on a site relative to the center of the density of states (DOS), zero in this case; and σ is the standard deviation of the distribution, which decides the amount of energy disorder. The Gaussian shape of the density of state is assumed as a Gaussian profile of the excitonic absorption band [6]. The approximation of the shape of the onsite energy distribution was confirmed experimentally using the direct measurement of the energy distribution in an electrochemically gated polymer transistor [26]. For materials in which the disorder aspect is dominant and the electron-phonon coupling is weak enough to induce the polaronic effects in the conduction process, although strong enough to ensure coupling to the heat bath, a sequence of each charge carrier transfer rate from one site to another is described by the Miller-Abrahams expression [27],

$$\nu_{ij} = \begin{cases} \nu_0 \exp\left(-\frac{2\mathbf{r}_{ij}}{\alpha} - \frac{\varepsilon_j - \varepsilon_i}{k_B T}\right) & \text{if } \varepsilon_j > \varepsilon_i, \\ \nu_0 \exp\left(-\frac{2\mathbf{r}_{ij}}{\alpha}\right) & \text{if } \varepsilon_j \le \varepsilon_i. \end{cases} \tag{2}$$

where $\mathbf{r}_{ij}$ is the distance between the positions of a charge carrier before and after hopping. The coefficient ν_0 is an intrinsic rate regarded as an attempt frequency; α is the localization length of the wave function of a charge carrier, which is assumed to be the same for sites i and j; k_B is the Boltzmann constant; T is the temperature; and ε_i and ε_j are the onsite energies of a charge carrier when at sites i and j, respectively. When there is an applied electric field, the corresponding electrostatic energies due to the field are added to ε_i and ε_j. The term $\exp\left(\frac{-2\mathbf{r}_{ij}}{\alpha}\right)$ is the overlap between the sites' wave functions, which decreases exponentially with the intersite distance $\mathbf{r}_{ij}$. The Boltzmann factor, $\exp\left(-\frac{\varepsilon_j-\varepsilon_i}{k_B T}\right)$, shows an activated process or the probability of the existence of a phonon of energy equal to $\varepsilon_j - \varepsilon_i$, which is either absorbed or emitted in order to conserve energy in the hopping process. There is no other activation energy, except the difference in charge carrier energies between different sites a carrier has to overcome in order to hop. The probability for hopping downward in energy is just $\nu_0 \exp\left(\frac{-2\mathbf{r}_{ij}}{\alpha}\right)$ by the principle of detailed balance with the premise that there are phonons that always exist to absorb the energy difference between the initial and final states. In ODSs, the intersite distance is also subjected to a variation in the intersite electronic wave function overlap that arises from both positional and orientational disorder of nonspherical molecules [28] known as off-diagonal disorder.

Bässler [6] utilized kinetic Monte Carlo (kMC) simulation techniques, incorporating the equations represented by Eqs. (1) and (2), to develop a comprehensive equation. This equation characterizes the charge carrier mobility as a function of temperature and electric field in ODSs. In this formulation, it was assumed that the onsite energies in the system are uncorrelated apart from the Gaussian shape of their distribution and also described the off-diagonal disorder, specifically the overlap parameter $\frac{2b}{\alpha}$, which was modeled as a Gaussian distribution with a specific width assigned to each site. This approach is known as the Gaussian disorder model (GDM). The Gaussian disorder model (GDM) equation describes temperature and electric field dependence for charge carrier mobility. According to this equation, the temperature dependence follows a non-Arrhenius relationship, given by $\mu \propto e^{\left(\frac{-2\hat{\sigma}}{3}\right)^2}$. The electric field dependence, on the other hand, adheres to a Poole-Frenkel mobility variation [29], represented as $\mu = \mu_0 e^{\sqrt{\frac{F}{F_0}}}$. However, this relationship is valid only within a limited range of electric fields, as experimentally demonstrated by Gill [13]. In the equations, $\hat{\sigma}$ is defined as the ratio of the standard deviation of the energy distribution (σ) to the product of the Boltzmann constant (k_B) and the temperature (T). The parameters μ_0 and F_0 are constant parameters that depend on the material and temperature. However, GDM has been widely used to analyze experimental data for transport in ODS [30, 31].

Identifying this inconsistency, ongoing efforts have been devoted to developing an equation that can effectively describe the Poole-Frenkel behavior across a wider range of electric fields [22, 32–37]. Throughout this undertaking, novel concepts have been suggested, mainly focusing on the spatial correlations among energy states across different sites. Analytical methods and kinetic Monte Carlo (kMC) simulations have been employed to determine and examine potential solutions. Gartstein and Conwell [34] were the first to propose a solution in this regard. They considered the spatial correlations between the energies of neighboring sites and performed Monte Carlo (MC) simulations. Their simulation results described the relationship between charge carrier mobility (u) and electric field (F) over a wide field range similar to experimental results [13]. Constructing a similar concept, another research group [35] developed an analytical framework to describe one-dimensional (1D) transport. They considered the influence of long-range energy correlations resulting from charge-dipole interactions within the material. By combining these correlations into their analytical model, they obtained results that aligned with the experimental observations. The concept of long-range energy correlations arising from charge-dipole interactions in the material was further expanded to confine three-dimensional (3D) motion. In this extension, the positional disorder was previously considered in Ref. [6], which was neglected. Subsequently, Novikov et al. [22] utilized Monte Carlo (MC) simulation results to suggest a phenomenological equation that empirically relates mobility to the electric field. The Correlated

Gaussian Disorder Model (CGDM), which was formulated for this particular model, has effectively shown the Poole-Frenkel behavior over a field range comparable to experimental observations [13]. This equation is particularly applicable to polar materials, as it was derived under the assumption of a randomly oriented dipole moment at each lattice site, with a large concentration assumption [33]. However, the Poole-Frenkel behavior has been observed in various conjugated polymers, including those that lack permanent dipole moments [18].

The studies mentioned earlier did not show significant emphasis on the influence of charge carrier density when investigating the relationship between charge carrier mobility, temperature, and electric field. However, the effects of charge carrier density on hole mobility in solution-processable conjugated polymers, specifically those developed for polymer light-emitting diodes (PLEDs) and polymer field-effect transistors (PFETs), have been reported in studies involving diodes and field-effect transistors [14, 15, 38]. In the experimental setup of space-charge limited polymer light-emitting diodes (PLEDs), it becomes challenging to distinguish the effects of charge carrier density and electric field on mobility. This difficulty arises because both the charge carrier density and electric field are simultaneously varied when adjusting the voltage across the thickness of the polymer film. However, during the analysis of experimental data, the variations in the electric field that occur with different voltages (particularly for voltages below 3 V) and, consequently, the dependence of mobility on electric field were neglected. The experimental data on mobility as a function of charge carrier density were gathered from unified diode and field-effect measurements and subsequently analyzed at various temperatures. The analysis was performed using the equation proposed by Vissenberg and Matters [39]. However, it is important to note that this equation only considers the dependence of mobility on charge carrier density while disregarding the field dependence of mobility. As a result, the equation fails to reproduce the experimental current versus voltage characteristics, particularly at low temperatures and high voltages. Recognizing the discrepancy between the previous equation and experimental observations, Pasveer *et al.* [16] developed a unified theoretical framework to describe charge carrier mobility. Their formulation incorporated the influences of temperature, charge carrier density, and electric field by numerically solving the master equation for charge carrier hopping on a regularly spaced lattice with uncorrelated energies following a Gaussian distribution. The numerical results obtained from this approach exhibited excellent agreement with experimental current-voltage data for two semiconducting polymers. This theoretical framework has been referred to as the extended Gaussian disorder model (EGDM).

The formulations presented in Refs. [6, 16] for the electric field dependence of hopping mobility have been extensively utilized for analyzing experimental data. However, recent literature has raised concerns about the ability of these formalisms to consistently and accurately describe the electric field dependence of hopping mobility [12, 40, 41]. The argument against the previous expressions is supported by the absence of a localization radius within each of these formulations. In systems consisting of randomly distributed localized states that can mimic organic disordered semiconductors (ODSs), the significance of this parameter in determining the dependence of mobility on the applied electric field cannot be overlooked. This has been revealed through computer simulations and analytical calculations in studies such as Refs. [12, 40, 41]. Another criticism, specifically directed toward the formalism presented in Ref. [16], is the omission of spatial disorder from the model and the utilization of a smaller localization length compared to Refs. [6, 30]. When a localization length of such magnitude is employed, charge carrier hopping becomes confined to the nearest-neighboring sites, resulting in a diminished electric field dependence on mobility. The dependence of charge carrier mobility on temperature and electric field was described in a study by Oelerich et al. [41]. In this work, an effective temperature incorporating the electric field was introduced to account for the combined influence of temperature and electric field on mobility. Additionally, research findings have provided evidence for the significance of both lattice site spacing and localization radius in determining the electric field dependence of mobility in regularly spaced sites. Conversely, in the case of randomly distributed sites, the localization radius has been identified as the only spatial parameter governing the field dependence of mobility. Based on these observations, the lattice models are inadequate for theoretical investigations of mobility versus electric field in spatially disordered systems [40].

Several previous studies [16, 22–24] have examined charge carrier mobility as a function of electric field, temperature, charge carrier density, energetic disorder, and spatial parameters using computer simulations. However, it is important to note that these studies did not consider the impact of spatial (off-diagonal) disorder on the mobility analysis. In contrast, there are Monte Carlo (MC) simulation methods [6, 9, 12, 32, 40, 41] that specifically consider spatial disorder in their approaches. As previously mentioned, in these simulations, the spatial disorder is modeled by assigning the overlap parameter $\frac{2b}{\alpha}$ to each site, following a Gaussian distribution with a variance Σ around the value $\frac{2b}{\alpha} = 10$ [6]. Recent literature, specifically studies by Refs. [12, 40, 41], have addressed the issue of spatial disorder by incorporating randomly distributed hopping sites in three dimensions. In these simulations, the average intersite distance is maintained at a constant value that corresponds to the inverse of the cube root of the site density.

In this chapter, we present a novel approach to incorporate spatial disorder and examine the relationship between charge carrier mobility and electric field, temperature, charge carrier density, energetic disorder, and localization radius using kinetic Monte Carlo (kMC) simulation methods. Our model is designed to localize a charge carrier within a sphere centered on regularly distributed sites. The radius of the sphere is directly related to the localization radius, allowing us to investigate the impact of this parameter on charge carrier mobility. This assumption is based on the understanding that charge carriers, such as electrons, are not strictly localized at a single point but instead occupy a certain region. The size of this region is determined by the width of the charge carrier's wave function. In our model, charge carrier hopping occurs between sites located randomly within different spheres. For example, the hopping distance from inside one sphere to a neighboring sphere can vary from $(b - 2R)$ to $(b + 2R)$, where b represents the distance between the centers of the neighboring spheres and R denotes the radius of a sphere. Using our proposed model and considering weak electron-phonon coupling, we incorporated uncorrelated Gaussian distributed energies and utilized the Miller-Abrahams rate equation to perform kMC simulations. Through these simulations, we obtained the necessary data for analysis. The obtained results are primarily focused on examining the relationship between mobility and charge carrier density, temperature, localization length, and electric field. The significance of the localization length in the correlation between mobility and other parameters is also discussed and analyzed. In Section 2, we describe the detailed procedures of our kMC simulation methods. The main numerical results and discussions are presented in Section 3. Finally, a summary and conclusions are given in Section 4.

2. Methodology

2.1 Kinetic Monte Carlo method

The kinetic Monte Carlo (KMC) method is a version of the Monte Carlo (MC) method designed to simulate the time evolution of specific processes. It is a powerful technique that allows simulation timescales to advance well beyond the vibrational timescale. The timescale between sequences of states is far beyond the vibrational timescale. Unlike the Metropolis MC method, the KMC method utilizes different approaches for generating the next state. The key concept behind KMC is to employ transition rates that depend on the energy barriers between states, with time increments developed to reflect the microscopic kinetics of the system. In the Metropolis MC method, the decision to accept a move is based on the energy difference between states. The KMC method employs rates that consider the energy barrier between states. KMC generates a sequence of configurations and the corresponding times at which transitions between these configurations occur. This approach effectively solves the master equation [42], as it provides a probability distribution for obtaining a certain configuration at time t, denoted as $p_i(t)$, which satisfies the master equation. The distinguishing feature of KMC is that it does not explicitly compute the probabilities $P_{ij}(t)$ but instead starts with a specific configuration that represents the initial state of the simulated experiment. It then generates a sequence of other configurations with the correct probabilities. There are various algorithms that are available that yield such a sequence of configurations.

There are stochastic algorithms available that accurately propagate a system from one state to another. The first algorithm, although valid, is less efficient and not widely used. In this approach, we assume that all rate constants for each state are known. In our system in state i, with a set of pathways and associated rate constants $\{k_{ij}\}$, we can determine the probability distribution $p_{ij} = k_{ij}\exp(-k_{ij}t)$ for each pathway. By using the relation $t = -\frac{\ln(r)}{k}$, we can generate an exponentially distributed time t_j for each pathway j. Among these pathways, only one will be the actual escape route. To determine this, we identify the pathway $j_{\min}$ with the smallest value of t_j and advance our system clock by $t_{j_{\min}}$. We then move the system to a new state and begin again from this new state. But, this is less than ideally efficient because we draw a random number for each possible escape path. The second algorithm, which is more commonly used and efficient, can propagate the system from one state to another using only two random numbers. The procedure begins by identifying all N possible processes, or viable escape pathways, from the current system configuration (state i). The number of pathways is limited to a finite value N, which corresponds to a maximum distance or maximum number of pathways for each particle within the time $t = -\frac{\ln(r)}{k}$. The N different rate constants are assumed to contribute to the total rate constant $k_{\text{tot}} = \sum_{p=1}^{N} k_p$. From these pathways, we select a process that satisfies a certain condition or criterion.

$$\frac{\sum_{p=0}^{q-1} k_p}{k_{tot}} \le r < \frac{\sum_{p=1}^{q} k_p}{k_{tot}}, \tag{3}$$

In our context, we have defined a condition where $\frac{k_o}{k_{\text{tot}}} = 0$, and r represents a random number drawn from a uniform distribution between 0 and 1. The expression in Eq. (3) implies that the rate constant k_p is added sequentially until reaching a step where $r < \frac{\sum_{p=1}^{q} k_p}{k_{\mathrm{tot}}}$ is satisfied, indicating the selected pathway. The maximum value of q is N, at which $\frac{\sum_{p=1}^{q} k_p}{k_{\mathrm{tot}}}$ becomes 1. This procedure ensures that the probability of choosing a specific pathway is proportional to its rate constant. To advance the system clock, we generate a random time from the exponential distribution associated with the total rate constant k_{tot}, using the expression $t = -\frac{\ln(\omega)}{k_{\mathrm{tot}}}$, where ω is another random number drawn from a uniform distribution between 0 and 1. It's important to note that this time is unrelated to the selection of events or pathways. The transition time for the system from one state to another solely depends on the overall escape rate. Once the system transitions to the new state, the list of pathways and their rates is updated, and the entire procedure is repeated.

2.2 Models

We are studying a system that performs as a model for a conjugated polymer film, similar to the one depicted in **Figure 1**. This film can be used as an active material in electronic devices like field-effect transistors (FETs), light-emitting diodes (LEDs), or photovoltaic cells. The geometry of this material can be assumed as a three-dimensional rectangular box with dimensions L_x, L_y, and L_z. Applying a voltage across the material between two metallic electrode contacts, an electric field is established. Charge carriers are injected into the polymer film from the electrode contacts, and under the influence of the electric field, they move from one contact (source) to the other (drain). The direction of motion of the charge carriers aligns with the electric field if they are holes, and it is opposite if they are electrons. In polymer light-emitting diodes, the electric field and the density of charge carriers depend on the potential difference between the electrodes. However, in polymer field-effect transistors, the charge carrier density is determined by the gate voltage, while the electric field along the channel length arises from the potential difference between the source and drain electrodes. In experimental setups, the values of these voltages and the geometric characteristics of the devices are known, allowing for the determination of the range in which the charge carrier density can vary for each type of device. It is found that the charge carrier density varies from $10^{14} cm^{-3}$ to $10^{16} cm^{-3}$ for PLEDs and from

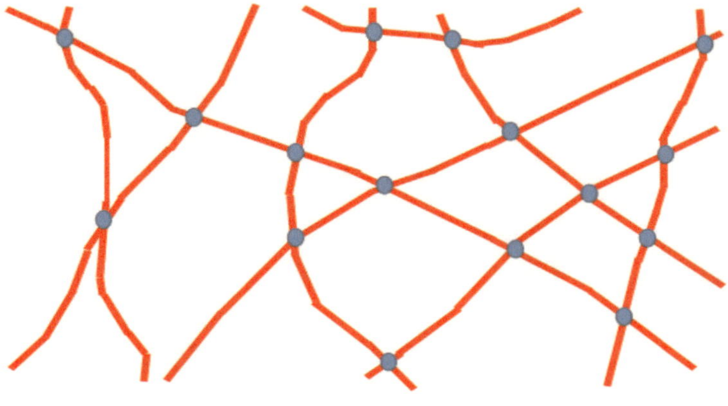

Figure 1.
Structure of thin film built from conjugated polymer chains, where the red and blue colors represent polymer segment and sites of the polymers, respectively [43].

$10^{17}cm^{-3}$ to $10^{19}cm^{-3}$ for PFETs [14, 15]. In this chapter, we assume similar devices and introduce a specific number of charge carriers into the polymer films, in such a way that the density of charge carriers exhibits the variation for both low and high densities. We do not investigate the dependence of charge carrier density on parameters or factors affecting it, as our focus is not on the effects of charge carrier density on mobility. The direction of the electric field between the source and drain electrodes, denoted as F_{SD}, is chosen to align with the $x - axis$, which also serves as the direction for calculating mobility. The magnitude of the electric field applied across the film is approximated based on the operational potential difference of a specific device arrangement. The width of the conducting channel is in the y-direction, while the $z - axis$ is perpendicular to both the y-axis and the channel length in the polymer film.

The model system we consider is a supercell of a hypothetical simple cubic lattice, with a lattice parameter denoted as b. Each subunit of a polymer chain is represented by a lattice point within this lattice. These lattice points, known as states or sites, are localized on different molecules or segments of the conjugated polymer. The length of each segment, which we assume corresponds to a lattice site, typically has an intrachain conjugation length of approximately $6 - 7nm$, and an interchain packing distance of around $0.3 - 0.4nm$. As a result, the distance between a site on one chain and its neighboring sites on the same chain differs from that on nearby chains. In a three-dimensional conjugated polymer system, the hopping process of charge carriers is anisotropic. Identifying precise microscopic parameters for a polymer film and using them accurately are challenging, as the film consists of a disordered arrangement of intertwined chains. Therefore, we approximate the hopping problem by considering the average behavior of both intrachain and interchain hopping processes, as done in other studies [6, 16, 22, 33]. The lattice parameter b is also treated as an effective result obtained by averaging the interchain and intrachain lattice spacings. A disordered system is created by randomly distributing $L_x \times L_y \times L_z$ lattice sites within a cubic supercell of side length L in the $x-$, $y-$, and $z-$ directions, respectively. Periodic boundary conditions are applied in all directions. The average intersite distance between neighboring sites is $b = N^{-1/3}$, where N represents the concentration of randomly distributed localized sites or states. The estimated value of the parameter N for organic conjugated polymers lies within the range of approximately $N \approx 10^{20}cm^{-3}$ to $N \approx 10^{21}cm^{-3}$ [44]. The sites involved in charge carrier hopping reside within a sphere centered at regularly placed points, as depicted in **Figure 2**, for sites distributed in two dimensions.

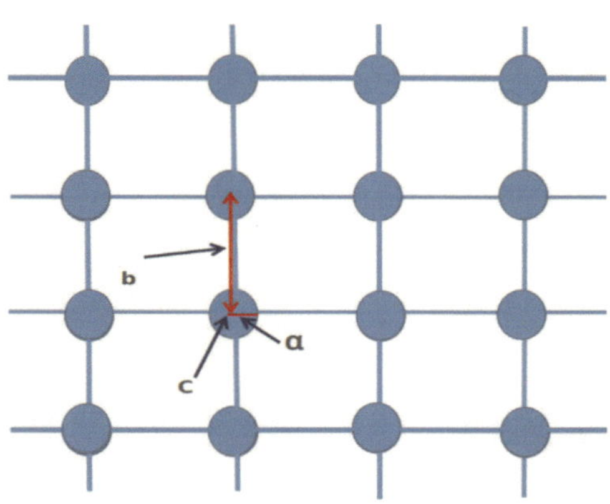

Figure 2.
Regularly spaced spheres of radius r, distance between centers of neighboring spheres b, and localization length α [43].

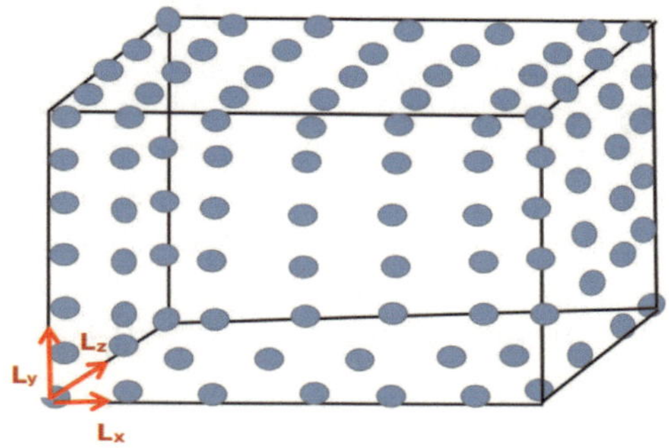

Figure 3.
Three-dimensional cubic supercell of lattice with sides L_x, L_y, and, L_z [43].

The distance between the centers of the nearest-neighboring spheres remains consistent and is equivalent to the lattice parameter denoted as b. However, the distance between the nearest-neighboring sites can range from $b - 2r$ to $b + 2r$, where r represents the radius of each sphere and is associated with the localization radius of a charge carrier's wave function. This implies that a charge carrier is placed randomly at any point inside a sphere, and the minimum hopping distance becomes a random variable that varies between $b - 2r$ and $b + 2r$. In situations where spatial disorder is not considered, r is assumed to be zero, and the sites are uniformly distributed on regularly spaced grids within a cubic box of side length L_x, as illustrated in **Figure 3**. To simulate energetic disorder in the system, we assign energies drawn from a Gaussian distribution described by Eq. (1) at the randomly distributed hopping sites.

2.3 Simulation details

As mentioned earlier, we consider a system with $L_x \times L_y \times L_z$ charge carrier hopping sites in three dimensions. The average number of sites is assumed to be the same along each dimension, and the average distance between neighboring sites, or the lattice spacing, is set to unity. The average number of sites per unit volume is $N = 10^{21} cm^{-3}$. A specific number N_h of charge carriers are introduced and randomly placed on different sites. The electric field along the x-axis, resulting from the source-drain bias V_{SD}, is assumed by its average value using the relation $E_{SD} = \frac{V_{SD}}{L}$, where L represents the length between the source and the drain electrodes. This assumption is based on the observation that this field remains nearly constant in the majority of the region where charge carriers move, as recently confirmed in Ref. [45]. Consequently, we can calculate the electric potential energy for a charge carrier at each regularly spaced site due to the source-drain bias field using the following equation.

$$U_{SD}(x, y, z) = F_{SD} e b (x - L_x) = \frac{V_{SD}}{L_x} b (x - L_x) \tag{4}$$

where the minimum is fixed at the right edge of the polymer film or at $x = L_x b$ in this case. Note that $x = 0$ and $x = L_x b$ coincide with the left and right edges of the polymer film, respectively.

Following the Gaussian disorder model (GDM), all sites where a charge carrier can hop to are assigned random energies drawn from a Gaussian distribution function centered at zero, as given in Eq. (1). The random energy assigned to a site represents the energy of a charge carrier when it occupies that particular site. Similar to the approach in Ref. [6], these energies are assumed to be uncorrelated. The energy of a charge carrier at a site is denoted as ε, and the standard deviation σ determines the level of energy disorder in the film. By mapping this Gaussian distribution to a uniform distribution of random numbers between 0 and 1, it becomes possible to determine the disorder energy for a charge carrier at each site in terms of the random numbers and a material-specific Gaussian width σ. The difference in energy between two hopping sites, denoted as i and j, can be described as follows.

$$U(\mathbf{R}_j) - U(\mathbf{R}_i) = [\varepsilon(\mathbf{R}_j) + U_{SD}(\mathbf{R}_j)] - [\varepsilon(\mathbf{R}_i) + U_{SD}(\mathbf{R}_i)], \tag{5}$$

The potential energy contribution from the source-drain field, denoted as U_{SD}, is determined by the electric field $\mathbf{F_{SD}}$. We have neglected the effect of electrostatic potential energy due to Coulomb interaction after independently verifying its negligible effects on the dynamics of charge carriers in the low density. Because, for densities lower than or equal to $10^{16} cm^{-3}$, the average distance between charge carriers is larger than 21 lattice parameters. Additionally, the static electric fields generated by the charge carriers have minimal influence on the transport properties within the lattice. This is due to the fact that the contribution of these charges to the energy difference between neighboring hopping sites, which determines the small step size and drift speed, is relatively negligible. In addition to this, we also assume that double occupancy of the same site does not occur as it is not energetically favorable for the level of the carrier density we are now dealing with and weak electron-phonon coupling. In the code, the rectangular coordinates of all the sites are known during the formation of the three-dimensional lattice. The charge carriers are assigned names from 1 to N_h when they are initially introduced to the sites. As a result, the coordinates of each charge carrier are known at all times in the code, and the magnitude of the distance between any two sites $\mathbf{R_{ij}}$ can be easily calculated numerically using the following relation.

$$R_{ij} = \sqrt{(x_i - x_j)^2 + (y_i - y_j)^2 + (z_i - z_j)^2}. \tag{6}$$

2.4 Simulation procedures

In our simulation, we begin by creating a three-dimensional lattice consisting of regularly spaced points or locations in a size of $L_x \times L_y \times L_z$. Each of these points represents the center of a sphere with a radius of R. We then randomly generate unoccupied sites within each sphere, which serve as hopping sites for charge carriers. Next, we compute the onsite disorder energies $\varepsilon(x, y, z)$ for each site using Eq. (5) and calculate $U_{SD}(x, y, z)$ using Eq. (4). This step involves assigning energy values to each site based on the given equations. In the third step, we introduce a specified number of charge carriers, denoted as N_h, into the system. These charge carriers are randomly placed on the unoccupied sites created in the first step. The value of N_h is chosen such that the charge carrier density falls within the desirable range for OLEDs or OFETs. For instance, for gate voltages V_g within the operating range of OFETs -10 V to -30 V, we use several charge carriers that correspond to densities ranging from $10^{17} cm^{-3}$ to $5 \times 10^{18} cm^{-3}$. Similarly, for source-drain voltages V_{SD} within the operating range of

OLEDs 1 V to 3 V, we use charge carrier numbers that correspond to densities ranging from $10^{14} cm^{-3}$ to $10^{17} cm^{-3}$. In the fourth step, we simulate the actual three-dimensional distribution of charge carriers based on the introduced energy values. Initially, the charge carriers are randomly distributed throughout the entire volume of the film. We assume that an equilibrium state of charge carrier distribution is achieved after a certain number of Monte Carlo time steps (MCTSs). To determine the hopping probabilities for each charge carrier, we follow a procedure where we examine the neighboring sites of each particle. If there are no empty sites nearby, we assign a probability of zero for the particle to hop and move on to the next particle. However, if there are neighboring unoccupied sites, we calculate the total energy difference between the particle's initial site and the neighboring empty sites using Eq. (5). We then solve for the transition rate, excluding ν_0, for each possible hopping site using Eq. (2). This procedure is repeated for all charge carriers, and the probability of hopping for each possible path is determined by dividing each transition rate by the total transition rates, as shown in Eq. (8). The sum of the probabilities for all possible paths is normalized to 1, and the transition from one configuration to another takes place by one of the N_h charge carriers along one path. This means that we use the probability length for each path and realize a transition of a system of charge carriers from one configuration to another by moving one of the charge carriers based on a stochastic approach.

The probability of a charge carrier transitioning from an occupied site i to an unoccupied site j can be expressed using Eq. (2), which represents the rate equation.

$$P_{ij} = \frac{\nu_{ij}}{\sum_{j=1}^{N_h} \sum_{i=1}^{l} \nu_{ij}}, \tag{7}$$

The values of ν_{ij}, obtained from Eq. (2), represent the transition rates between occupied site i and unoccupied site j. In this equation, i enumerates the possible nearest-neighboring sites for each charge carrier, ranging from one to say, l. Similarly, j counts the number of charge carriers, ranging from one to the last, denoted as N_h. Eq. (7) ensures that the system satisfies a detailed balance condition in the steady state. We need to assign probabilities to each possible destination site. These probabilities form a sequence of intervals between 0 and 1. The sum of all these probabilities, as given by $\sum_{i=1}^{N_h} \sum_{j=1}^{l} P_{ij}$, equals 1. To choose a particular hop, we generate a random number r from a uniform distribution within the interval of 0 to 1. This random number corresponds to a specific interval and consequently, identifies the destination site to which one of the charge carriers will jump. The selection process follows a scheme similar to that defined in Eq. (3).

$$\sum_{i=0}^{N_h-1} \sum_{j=0}^{l-1} P_{ij} \leq r < \sum_{i=1}^{N_h} \sum_{j=1}^{l} P_{ij}, \tag{8}$$

We assign the value of P_{00} as 0. In Eq. (7), we limit the potential destination sites j to the 26 nearest-neighboring sites within an average volume of a $3 \times 3 \times 3$ lattice. This restriction is imposed because, even with a larger number of possible destinations, the majority of hopping events primarily occur between nearest-neighboring sites. As a result, the impact of this restriction is negligible. This process is repeated for a specific number of Monte Carlo time steps (MCTSs), until we record enough data.

Each MCTS involves moving one charge carrier after selecting from all charge carriers, according to the procedure described above. In the context of the conjugated polymers described in Refs. [6, 46], the average time a charge carrier spends at a specific site i, known as the mean dwelling time τ, is equivalent to the average time for a transition to occur between the states identified through the previous procedure is given as,

$$\tau = \frac{1}{\sum_j \nu_{ij}}. \tag{9}$$

$$\tau = \frac{1}{\sum_{j=1}^{N_p} \sum_{i=1}^{26} \nu_{ij}} \tag{10}$$

The actual hopping time, denoted as τ_h, follows a distribution around the mean time τ. The range of possible hopping times extends from near-zero to infinity. This distribution can be described by an exponential probability distribution of the form $\frac{1}{\tau}\exp\left(-\frac{\tau_h}{\tau}\right)$, as discussed in the preceding section. To determine the specific value of the actual hopping time τ_h, we can consider

$$p(\tau_h) = \int_0^{\tau_h} \frac{1}{\tau} \exp\left(-\frac{t}{\tau}\right) dt. \tag{11}$$

Given that the hopping time τ_h lies within the interval $[0, \infty)$, the probability distribution function $p(\tau_h)$ takes the value of 0 when $\tau_h = 0$ and 1 when $\tau_h = \infty$. For any other value of τ_h within this interval, the probability lies between 0 and 1. Consequently, we can map a random number r drawn from a uniform distribution between 0 and 1 to the value of τ_h. By evaluating the integral in Eq. (11) and solving for τ_h in terms of r and τ, we can determine the specific value of τ_h.

$$\tau_h = -\tau \ln(1 - r). \tag{12}$$

By incorporating stochastic behavior in both the transition path, as described in Eq. (7), and the transition time, the approach ensures that the average transition time between two configurations or states is equal to the mean of the inverse transition rates between those states.

To ensure that the system reaches a steady-state distribution, we allow the system to undergo $(m - 1) \times 10^n$ Monte Carlo time steps (MCTSs). It is important to note that slight but noticeable variations in the simulation results can arise due to different generations of onsite energies obtained from the Gaussian disorder model (GDM). Therefore, the results are averaged over 15 different generations of energy distributions to account for these variations.

The fifth and final step of the simulation involves recording the displacement of the charge carriers along the length in the direction of $\mathbf{F}_{SD}$ (the source-drain bias) and the corresponding time it takes for this displacement to occur. This recording is carried out during the last 1×10^n Monte Carlo time steps. By analyzing the displacement of charge carriers over time, we can determine their velocity. The mobility of the charge carriers is calculated by dividing the velocity and the source-drain bias electric field $\mathbf{F}_{SD}$. This process provides detailed information about the transport properties of the charge carriers. The simulation data are collected while varying the charge carrier concentration, source-drain voltage (F_{SD}), and temperature (T) for different localization radii.

3. Charge transport properties in disordered organic semiconductors: Monte Carlo simulation

This chapter primarily focuses on investigating the transport properties of charge carriers in organic disordered semiconducting-conjugated polymers. The main objective is to examine how the charge carrier mobility depends on various parameters characterizing the materials and externally applied parameters. These investigations are based on the results obtained from kinetic Monte Carlo simulations. One important parameter influencing the charge carrier mobility is the energy disorder strength, which is typically described by the standard deviation (σ) of the energy distribution. In our simulations, as done in previous studies [6, 16], the energies of charge carriers resulting from electric fields and temperature are expressed in terms of this disorder energy. Another parameter affecting charge carrier mobility is the localization length of the charge carrier wave function, which is expressed in terms of a lattice parameter known as the lattice site spacing (b), as explained in Section 2. In addition to these parameters, charge carrier mobility also depends on the charge carrier density (ρ) and the concentration of lattice sites (N). In our calculations, we consider a cubic sample with a volume of $200nm \times 200nm \times 200nm = 8 \times 10^6 nm^3$ containing 8×10^6 sites, resulting in an average of $N = 10^{21} cm^{-3}$ states or one site per nm^3. The lattice spacing is related to the site concentration as $b = N^{-1/3}$. We keep the total number of sites fixed and vary the charge carrier density (ρ) by adjusting the number of introduced charge carriers in the sample. The charge carrier mobility, which is the main transport parameter, is primarily expressed in terms of the lattice parameter (b), disorder parameter (σ), electron charge (e), and phonon frequency (ν_0), or units of $\frac{b^2 \nu_0 e}{\sigma}$. When these parameters are explicitly known, the charge carrier mobility can be expressed explicitly as well. The electric field magnitudes used in our calculations are chosen to cover different transport regimes, with most experimental data falling within the low field strength regime (below or equal to $1 \times 10^5 Vcm^{-1}$) [13–16, 18–20]. The numerical results obtained from the kinetic Monte Carlo simulations are presented and analyzed in the subsequent sections. In the first section, we present results for regularly spaced states without spatial disorder, where the movement of a random walker (or charge carrier) occurs within regularly spaced sites. In the second section, we include the effects of spatial disorder by considering the radii of spheres centered at regularly spaced positions. The state of a charge carrier is localized within a sphere, with only one state present in each sphere. The distance between neighboring spheres is the same as the lattice spacing (b), while the hopping distance from one sphere to another nearest-neighbor sphere ranges from $(b - 2R)$ to $(b + 2R)$, where R represents the radius of a sphere. This arrangement incorporates both positional and orientational disorder, with the random spatial disorder generated from a uniform distribution. This model simulates a film composed of polymer chains irregularly spaced and randomly oriented within a three-dimensional rectangular box, where each chain segment resides within a sphere representing an individual hopping site.

3.1 The effect of lattice site spacing and localization length in spatial disorder system for $\mu(F)$

In this section, we present the results of our simulations and focus on the impact of the disordered energy parameter $\hat{\sigma}$ on the relationship between charge carrier mobility and electric field strength. We consider different values of the localization length and spatial disorder parameter, denoted by the dimensionless quantity $r = \frac{R}{b}$. The simulations are conducted at room temperature ($T = 300K$) using energetic disorder parameters $\hat{\sigma} = 3$ and $\hat{\sigma} = 4$, corresponding to energy disorder values of $\sigma = 0.075eV$ and $0.1eV$, respectively. These values fall within the range of σ typically observed in organic disordered semiconductors (ODSs), which lies between 0.05 and $0.14eV$ [6, 16, 47–49]. **Figure 3** illustrates the curves depicting a decrease in mobility with increasing spatial disorder for both $\hat{\sigma} = 3$ and $\hat{\sigma} = 4$. Furthermore, we observe that the mobility increases with higher electric field strengths for r values less than or equal to 0.15, with a more significant effect observed for $\hat{\sigma} = 4$. However, for larger values of r, there is a decrease in mobility as the electric field strength increases. These results indicate that the impact of electrostatic potential energy variations on charge carrier hopping in the direction of the electric field is not uniform and becomes less influential than the effects of thermal energy when spatial disorder is more pronounced. This idea can explain the weaker dependence of mobility on the electric field observed in panel (a) compared to panel (b). In panel (a), for $\hat{\sigma} = 3$ and $r = 0.2$, the charge carrier mobility remains constant regardless of the electric field, while for $r = 0.25$, the mobility decreases with increasing electric field. This behavior aligns with what has been reported in recent literature [40, 41]. In panel (b), for $\hat{\sigma} = 4$ and $r = 0.2$ and $r = 0.25$, the charge carrier mobility also shows no variation with the electric field within a certain range, consistent with the findings in Refs. [40, 41]. These results confirm the correctness of our model and program by comparing them to previous results. However, the simulation data plotted in **Figure 4** for randomly placed sites with $\alpha = 0.1b$ and $r > 0.15$ do not align with experimental results [13, 51–53].

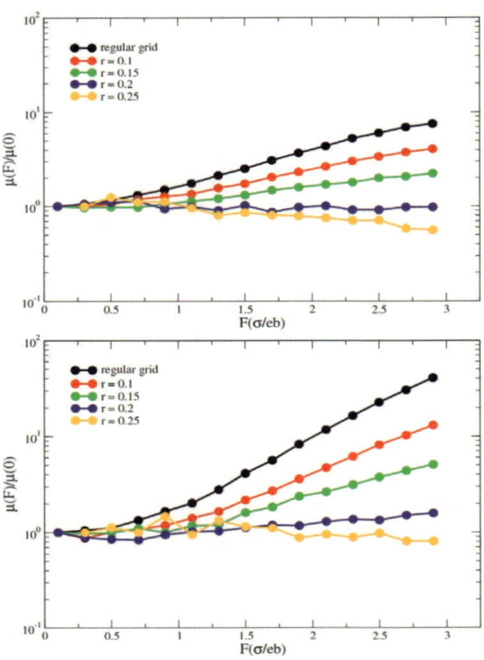

Figure 4.
Logarithm of the normalized charge carrier mobility as a function of the electric field for different values of spatial disorder parameter r, where e is the unit charge. Localization length $\alpha = 0.1b$, charge carrier density $p = 1 \times 10^{16} cm^{-3}$, and spatial disorder parameter r are in the range from 0 to 0.25, where $r = 0$ specifies a regularly spaced site. (a) $\hat{\sigma} = 3$ and (b) $\hat{\sigma} = 4$ [50].

Experimental studies indicate that the charge carrier mobility in ODS increases with the electric field. In contrast, our data for regularly spaced sites and spatial disorder parameters $r = 0.1$ and $r = 0.15$ demonstrate an increase in charge carrier mobility with the electric field until reaching a saturation value, consistent with experimental observations [13, 51–53], as well as computer simulations that did not involve spatial disorder [16]. The discrepancy between our simulation results and the experimental and previous simulation results [wf05] when $r > 0.15$ and $\alpha = 0.1b$ can be attributed to the chosen values of r compared to α. This indicates that the assigned value for the localization length ($\alpha = 0.1b$) for $r > 0.15$ may not be realistic in an experimental context. It is possible that either the considered localization length is smaller than the actual value [40, 41] or the spatial disorder parameter $r > 0.15$ when $\alpha = 0.1b$ is not reasonable. In general, for larger values of r, the weak dependence of charge carrier mobility on the electric field is primarily due to the limited impact of electrostatic energy compared to thermal activation when the distribution of localized states is random. Consequently, the electric field dependence of charge carrier mobility in regular grids differs from that in randomly spaced localized sites.

Figures 4 and **5** present logarithmic plots of charge carrier mobilities as a function of electric field for different values of the spatial disorder parameter r. The range of r spans from 0 to 0.25, where $r = 0$ corresponds to a regular grid. The simulations consider two disordered energy parameters, $\hat{\sigma} = 3$ and $\hat{\sigma} = 4$. In both **Figures 4** and **5**, an increase in charge carrier mobility with electric field is observed when the localization length α is $0.1b$ or $0.14b$, and r is either 0.1 or 0.15. However, for (4) with $\hat{\sigma} = 3$, when α is $0.1b$ or $0.14b$, the charge carrier mobility either decreases or remains unchanged with increasing electric field if r is 0.2 or 0.25. In contrast, **Figure 6** with

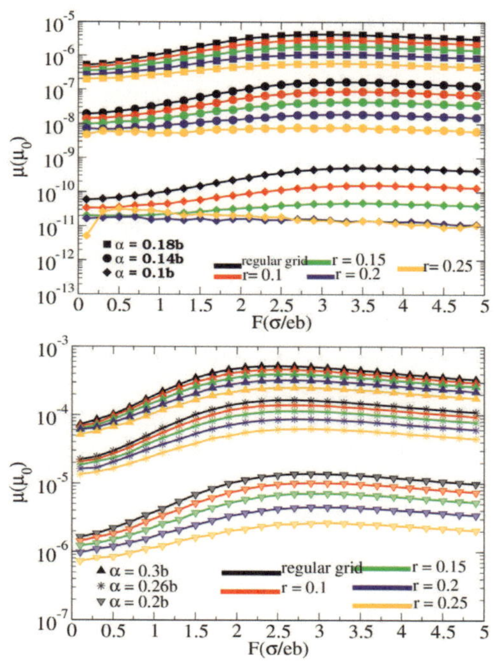

Figure 5.
Logarithm of the normalized charge carrier mobility as a function of the electric field for different spatial disorder parameters r in the range from 0.1 to 0.25 for the same disorder energy parameter $\hat{\sigma} = 3$ *and charge carrier density* $p = 1 \times 10^{16} cm^{-3}$ *where* $\mu_0 = \frac{b^2 \nu_0 e}{\sigma}$ *and for both regular grids and randomly spaced sites with different values of α in the range from (a) 0.1b to 0.18b and (b) 0.2b to 0.3b. [50].*

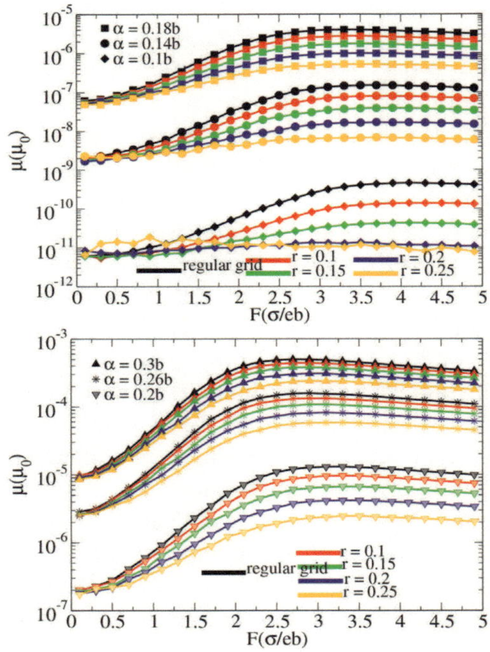

Figure 6.
Logarithm of the normalized charge carrier mobility as a function of the electric field for different spatial disorder parameters r in the range from 0.1 to 0.25 for the same disorder energy parameter $\hat{\sigma} = 4$ *and charge carrier density* $p = 1 \times 10^{16} cm^{-3}$ *where* $\mu_o = \frac{b^2 \nu_0 e}{\sigma}$ *and for both regular grids and randomly spaced sites with different values of* α *in the range from (a) 0.1b to 0.18b and (b) 0.2b to 0.3b [50].*

$\hat{\sigma} = 4$ and $\alpha = 0.1b$ demonstrates that the charge carrier mobility does not vary with electric field for r values of 0.2 and 0.25. However, when α increases to $0.14b$, the charge carrier mobility increases with the electric field. Overall, the increase in charge carrier mobility with electric field is more pronounced when α/b exceeds r and is within a certain range. Conversely, when α/b is smaller than r, the charge carrier mobility either decreases or remains unaffected by the electric field, as illustrated in both figures. Notably, the decrease in mobility with increasing electric field occurs at lower electric field values for $\hat{\sigma} = 3$. However, when $\hat{\sigma} = 4$ and α/b is smaller than r, the mobility does not change with the increase in the electric field within the mentioned range. These findings align with previous literature [41] that also reported similar observations, although without specifying the values of r and α/b. These peculiar results indicate a dependence of charge carrier mobility on the localization length and electric field. They also suggest that either the chosen α value is too small or the selected r value is unrealistically large compared to α/b. These findings support the notion proposed in Refs. [40, 41] regarding the importance of considering these two parameters in the devised model for studying charge carrier mobility behavior in organic disordered semiconductors (ODSs).

Based on our analysis of the presented results, it seems that realistic values for the spatial disorder parameter follow the condition $r \leq 2\alpha/b$. In **Figure 7(a)** and **(b)**, we have plotted the data generated to study the charge carrier mobility as a function of the electric field considering different values of r. For the specific values of r mentioned, we observe that the charge carrier mobility remains constant in a lower range of electric fields, but increases with the electric field in a higher range. Additionally,

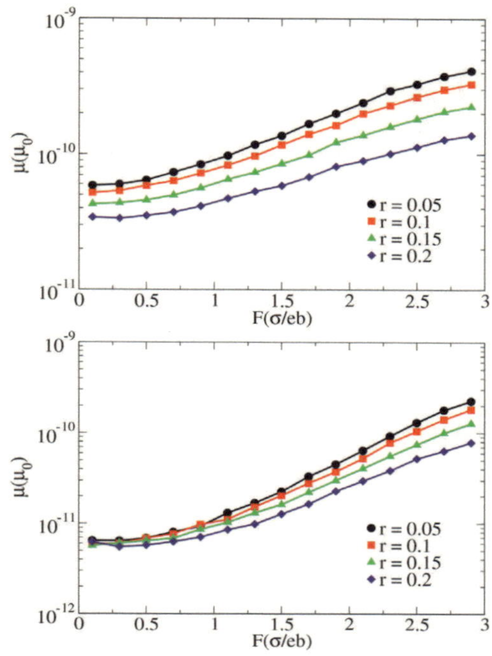

Figure 7.
Logarithm of the normalized charge carrier mobility as a function of the electric field for randomly distributed sites with different values of spatial disorder parameter r for (a) $\hat{\sigma} = 3$ and (b) $\hat{\sigma} = 4$ for the same localization length $\alpha = 0.1b$ and charge carrier density $\rho = 1 \times 10^{16} cm^{-3}$ for both figures, where b is lattice spacing and $\mu_o = \frac{b^2 \nu_0 e}{\sigma}$ [50].

we notice a significant variation in charge carrier mobility with the spatial disorder parameter r in the lower electric field region, particularly in panel (a). However, in panel (b), we only observe slight changes in mobility with r when the electric field is in the higher range. In **Figure 4b**, when $\hat{\sigma} = 4$ and $r = 0.2$, we observe similar independence of charge carrier mobility from the electric field in the lower electric field region. This characteristic, the independence of mobility from the electric field at $\hat{\sigma} = 4$, is not observed within the same electric field range for regularly spaced sites ($r = 0$), as demonstrated in this study and previous research [16]. However, at higher electric field values, we observe an increase in charge carrier mobility with the electric field when $\hat{\sigma} = 4$, as depicted in **Figure 7b**. This behavior is similar to what has been observed for regularly spaced sites ($r = 0$) and has also been reported in previous studies [15]. These findings imply that the data obtained by taking the spatial disorder parameter into account show a closer alignment with experimental results [14, 15] compared to the data generated for regularly spaced sites with $r \leq 2\alpha/b$. This suggests that the transport properties of charge carriers in organic disordered semiconductors (ODSCs) can be more accurately described by a lattice model where the localized states are randomly distributed, rather than arranged in a regular manner.

Figure 8 presents the simulation results depicting the relationship between charge carrier mobility and electric field for various values of the spatial disorder parameter $r \leq \frac{2\alpha}{b}$ and localization length $\alpha = 0.1b$. The figure also includes four different values of $\hat{\sigma}$ ranging from 3 to 6. By examining the figure, we can clearly observe the independence of charge carrier mobility from the electric field in the lower field region, with the extent of this effect increasing as $\hat{\sigma}$ becomes larger. However, in the higher field region, we notice an increase in charge carrier mobility with the electric field,

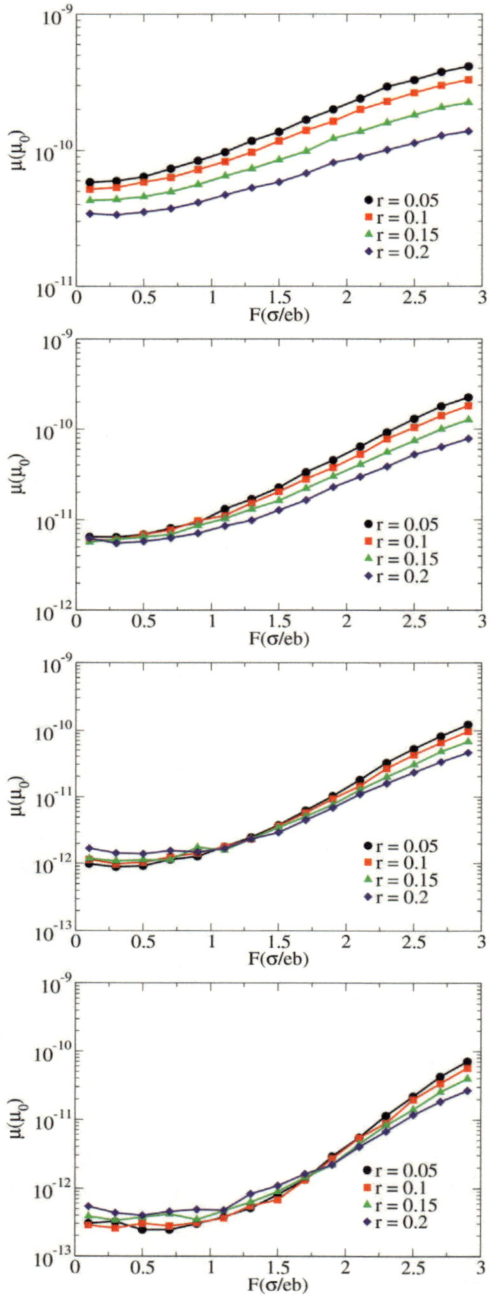

Figure 8.
Logarithm of the normalized charge carrier mobility as a function of the electric field for the same localization length $\alpha = 0.1b$, charge carrier density $\rho = 1 \times 10^{16} cm^{-3}$, where $\mu_o = \frac{b^2 \nu_o e}{\sigma}$ and different values of spatial disordered parameter (a) $\hat{\sigma} = 3$, (b) $\hat{\sigma} = 4$, (c) $\hat{\sigma} = 5$, and (d) $\hat{\sigma} = 6$ [50].

particularly for higher values of $\hat{\sigma}$. Notably, the curves corresponding to the largest spatial disorder parameter ($r = 0.2$) demonstrate that the mobility is the highest among all in the lower electric field region, while it is the lowest among all in the higher electric field region for $\hat{\sigma} = 5, 6$, as depicted in panels (c) and (d). These observations support the suggestion made in Ref. [15] that the influence of the electric field on charge carrier motion in the low field region is less significant compared to thermal activation when the hopping sites are randomly spaced. Nonetheless, in the larger field region, it appears that the electric field plays a more dominant role in the hopping motion of charge carriers, although its influence is relatively less pronounced for larger values of r, as indicated by the curves in panels (c) and (d).

3.2 The effect of spatial disorder on charge carrier mobility for charge transport

Figure 9(a)-(d) display simulation results of the normalized charge carrier mobility, which is the ratio of the mobility at various electric fields to the mobility at the minimum electric field. The electric field is expressed in units of $\frac{\sigma}{e\alpha}$. These plots explore the behavior of the normalized mobility as a function of the electric field for different combinations of localization length and spatial disorder parameters. **Figure 9a** and **b** specifically show the plots for $r = 0.25$ and $r = 0.2$, respectively. In these cases, the charge carrier mobility curves converge into a single curve, indicating that the variation in mobility with the electric field is consistent, regardless of the differences in the localization length. This convergence occurs when the electric field is expressed in units of $\frac{\sigma}{e\alpha}$, instead of $\frac{\sigma}{eb}$. However, for $r = 0.15$ and $r = 0.1$, as depicted in **Figure 9c** and **d**, the charge carrier mobility variations with electric field differ for different values of the localization length. These differences can be attributed to the magnitude of the spatial disorder parameter, r. When r is smaller, the normalized charge carrier mobility increases, with the electric field until reaching a saturation value corresponding to each localization length α. This behavior is evident in **Figure 9 (c)** and **(d)**. In contrast, for lower electric fields and larger spatial disorder parameters r, the charge carrier mobilities converge into a single curve and then separate and increase (with a faster rate for larger localization lengths) as the electric field increases until a saturation value is reached, corresponding to each localization length α. It is also worth noting that the saturation values for the same disorder energy parameter $\hat{\sigma}$ are not reached at the same electric field. If the localization length α is smaller, the saturation occurs at a lower electric field.

We have observed that the behavior of charge carrier mobility in response to an electric field depends on the values of the spatial disorder parameter r relative to α/b. Specifically, when r exceeds $2\alpha/b$, the charge carrier mobility either remains constant or decreases with increasing electric field. This implies that the electric field dependence of charge carrier mobility is primarily determined by the localization length and lattice site spacing (represented by b), which are crucial length scales. Additionally, the spatial disorder parameter r also influences the transport properties of charge carriers in ODSs. However, when the disorder parameter r surpasses $2\alpha/b$, it is the localization length alone that becomes the decisive spatial parameter governing the electric field dependence of charge carrier mobility, rather than the lattice site spacing. This behavior is consistent with the findings from kinetic Monte Carlo (kMC) simulation results reported in Refs. [40, 41].

3.3 The effects of disorder energy, charge carrier density, and electric field on carrier mobility

The curves displayed in both panels of **Figure 10** demonstrate that the mobility in the lower electric field region decreases as $\hat{\sigma}$ increases. This observation is consistent

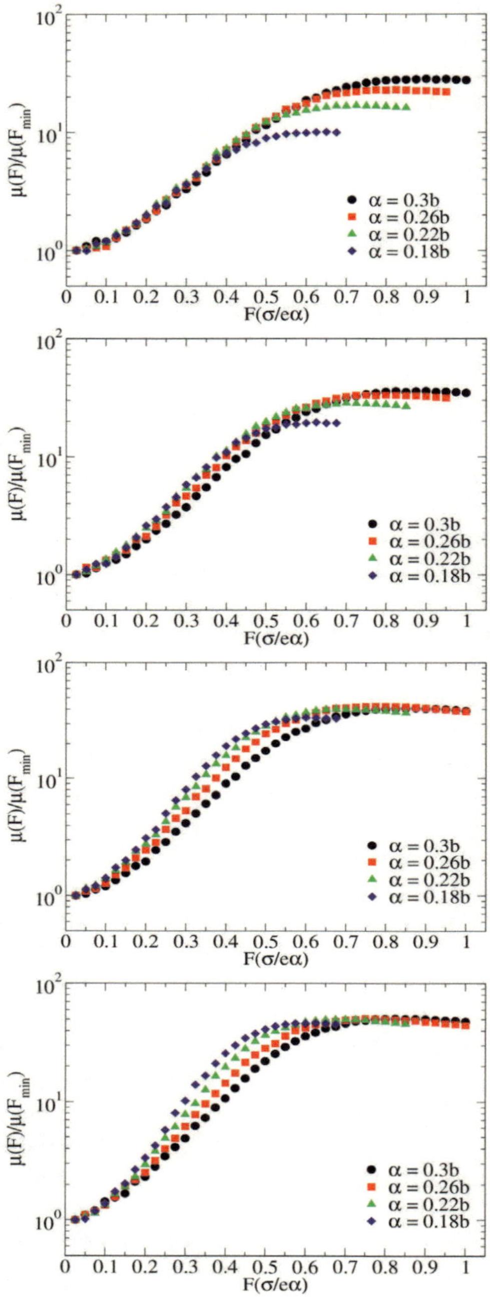

Figure 9.
Logarithm of the normalized charge carrier mobility as a function of the electric field for the same disorder energy parameter of $\hat{\sigma} = 4$ *and charge carrier density* $\rho = 1 \times 10^{16} cm^{-3}$ *and for different values of localization length and disorder parameter r (a)* $r = 0.25$*, (b)* $r = 0.2$*, (c)* $r = 0.15$*, and (d)* $r = 0.1$ *[50].*

with previous reports [6, 14–16] and supports their justifications. In this lower electric field region, the mobility is more influenced by temperature rather than the electric field itself. Transport primarily occurs through diffusion, and as a result, mobility remains relatively constant. However, for electric fields larger than $\frac{\sigma}{eb}$, the impact of

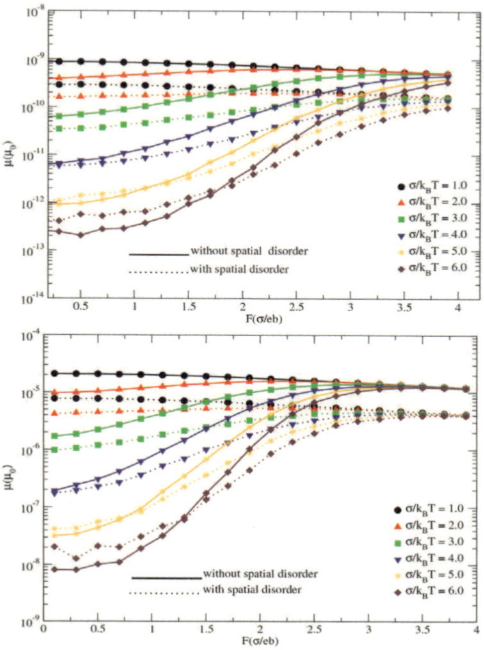

Figure 10.
Logarithm of the normalized charge carrier mobility as a function of the electric field for different values of $\hat{\sigma}$ and α and for the same density ($p = 1 \times 10^{16} cm^{-3}$). (a) $\alpha = 0.1b$, (b) $\alpha = 0.2b$, where b is a lattice parameter, and $\mu_0 = \frac{b^2 \nu_0 e}{\sigma}$ [54].

activation barriers on charge carrier hopping diminishes with increasing electric field. In this regime, conduction is predominantly driven by the electric field-induced drift, leading to an increase in mobility until a saturation value is reached for each $\hat{\sigma}$ greater than 1. Importantly, in the case of a smaller localization length, specifically when $\alpha = 0.1b$, the mobility exhibits an increasing trend with the electric field, until it eventually reaches a saturation point. This change becomes more pronounced as $\hat{\sigma}$ increases. The saturation phenomenon occurs because the electric field reduces the barrier height for uphill jumps of charge carriers in the direction of the field, thereby enhancing mobility in hopping conduction, as described in Ref. [29]. The drift mobility is calculated using the relation $\mu = v/F$, where v is the drift velocity of a charge carrier due to the electric field (F). Consequently, increasing the electric field is more effective in reducing the activation barrier, although it also reduces mobility, until the electric potential energy difference between any two sites separated by a distance x along the electric field, eFx, approaches (but does not exceed) the energy scale (σ). If eFx surpasses σ, the excess energy $eFx - \sigma$ does not contribute to speeding up hopping conduction but instead reduces mobility since it appears in the mobility calculation, $\mu = v/F$. In this context, eFx can be interpreted as an energy that lowers the activation barrier, and the increase in mobility with the electric field is significant as long as the difference between eFx and the average activation barrier height between hopping sites separated by a distance x is substantial. The reference energy value we used here, σ, to explain the statement that the electric field increases mobility until eFx reaches σ may not be exact and requires experimental verification, although it is reasonable and makes sense. Similarly, temperature plays a crucial role in reducing the difference between the energies at the transport level and the equilibrium level of the hopping particle, as discussed elsewhere [55, 56]. Most of the solid (or broken) curves in **Figure 10** exhibit similar overall behavior, except for an increase in mobility with an increase in localization length. Additionally, the saturation of mobility occurs at a lower electric field for larger localization lengths. This suggests that as the localization length increases, charge carrier conduction becomes more prominent, and the influence of the electric field on mobility decreases. One potential explanation for this phenomenon is that as α increases larger, the probability of charge carriers hopping to sites located farther away also increases. This leads to an increase in eFx, thereby enhancing the role of the electric field in reducing the barrier height between hopping sites separated by a distance x. Consequently, mobility saturation is observed at a relatively lower electric field when the localization length is larger compared to when the localization length is smaller. When the localization length is small (or the state is strongly localized), charge carrier hopping occurs rarely at relatively high $\hat{\sigma}$ in the lower electric field region. Another important inference from our simulation results is that the possible value of the localization length is closer to $0.2b$ than to $0.1b$. This suggestion, although it requires experimental verification, is supported by the fact that the observed mobilities are of the same order as those demonstrated experimentally [14, 15] when the localization length is $0.2b$, which is higher than the mobility observed for $\alpha = 0.1b$, which is much lower.

Moreover, when examining the solid and broken curves for $\hat{\sigma}$ values exceeding 2, a more pronounced fluctuation in mobility with the electric field is evident in the solid curves compared to the broken curves. Furthermore, it is generally observed that the mobility is greater when charge carriers hop on regularly distributed sites as opposed to randomly distributed sites, except in the low field region when $\hat{\sigma}$ is below 4. The plotted curves in **Figure 10**, specifically in the first panel, exhibit comparable patterns to those presented in [57] for randomly distributed sites featuring the same reduced disorder energy $\hat{\sigma}$ and localization length $\alpha = 0.1b$. In the region of low electric fields for $\hat{\sigma} = 4, 5$, and 6, each solid curve is positioned below its corresponding broken curve, and we notice that they intersect at specific electric field values. This observation suggests that the electric field exerts a more significant influence on mobility when it comes to hopping conduction on regularly distributed sites compared to randomly distributed sites. The reason for observing a solid curve above the broken curve for the same $\hat{\sigma}$ suggests that spatial disorder adds additional hindrance to the hopping conduction of charge carriers, especially for lower $\hat{\sigma}$ values. When $\hat{\sigma}$ is large and the electric field is low, a favorable destination site for a charge carrier is a state with lower energy located at a closer distance. In the case of hopping on randomly distributed states, the distance between neighboring sites varies from $(b - 2R)$ to $(b + 2R)$, and there is a probability of finding a favorable destination state at a distance less than the mean value (or lattice parameter), which is absent in hopping on regularly spaced states. This can explain why a solid curve lies below the broken curve for larger $\hat{\sigma}$ values in the low field region. When $\hat{\sigma}$ is small, the activation barrier is low, and a charge carrier can also hop to sites at larger distances. However, since there are nearby sites, the probability of hopping to nearby sites is higher, and as a result, a charge carrier that can hop to a site at a larger distance chooses to jump to a site at a shorter distance when hopping on randomly distributed sites compared to regularly distributed sites. Similarly, the presence of sites at distances less than the mean value reduces the influence of the electric field on charge carrier hopping in randomly distributed sites compared to equally spaced sites with the same site spacing.

The curves presented in **Figure 11** demonstrate a similar trend in mobility variation with the electric field as the corresponding curves in **Figure 10**, regardless of the

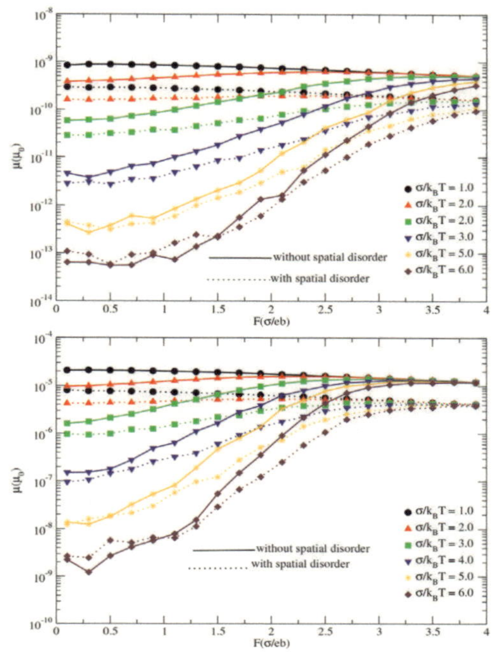

Figure 11.
Logarithm of the normalized charge carrier mobility as a function of the electric field for different values of $\hat{\sigma}$ and α and for the same charge carrier density ($p = 1 \times 10^{15} cm^{-3}$). (a) $\alpha = 0.1b$, (b) $\alpha = 0.2b$, where b is a lattice spacing, and $\mu_0 = \frac{b^2 \nu_0 e}{\sigma}$ [54].

charge carrier density values. Similar to **Figure 10**, the mobility increases with the electric field until it reaches a saturation value for $\hat{\sigma}$ values greater than 1. The increase in mobility becomes more pronounced as $\hat{\sigma}$ increases. Additionally, we observe an increase in mobility with charge carrier density, although the density is relatively low (within the range of densities for light-emitting diodes [14, 15]). The reason for the increase in mobility with charge carrier density can be attributed to the larger number of deep states, as suggested in prior works [57], that become filled with charge carriers as the density increases. Consequently, more states with closely lying energies become available for charge carrier hopping when the density becomes higher. In both **Figures 10** and **11**, we notice that the effect of electric field on mobility becomes stronger as $\hat{\sigma}$ increases, except in the low electric field region. In this low field region, we observe that mobility is independent of the electric field in **Figures 10(b)** and **11 (a)** when $\hat{\sigma} = 6$. The reason for not observing this behavior clearly in **Figure 11(b)** may be attributed to differences in density and localization length. The dependence of mobility on the electric field decreases with density, as suggested in Ref. [15]. The localization length in **Figure 11(b)** is larger than that in **Figure 11(a)**, and since the electrostatic energy *eFx* is larger for a larger localization length, we only observe changes in mobility with the field for larger localization lengths. The results obtained for larger values of $\hat{\sigma}$ exhibit similarities with those shown in Ref. [16], although there is a difference for $\hat{\sigma} = 6$ in the low-density region, which bears similarity to what is suggested in Ref. [15]. To further understand the slight deviations from experimental results, we have calculated mobility as a function of charge carrier density for different temperatures while keeping the electric field constant. We will present and discuss these results further below.

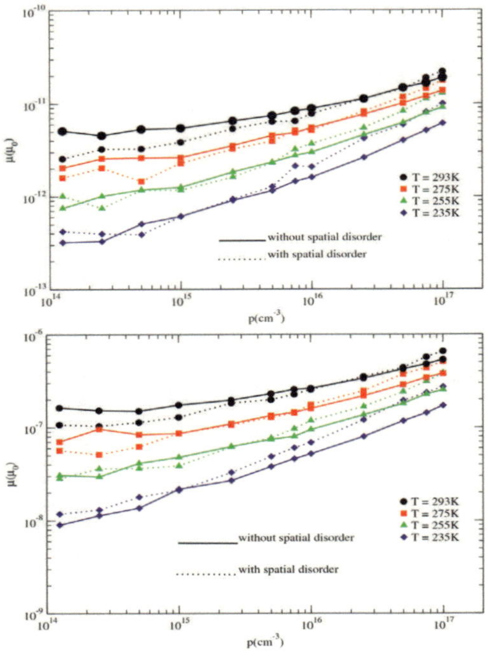

Figure 12.
Logarithm of the charge carrier mobility as a function of the charge carrier density for the same disorder energy $\sigma = 0.1eV$, at different temperatures (T), for the same electric field $F = 0.1 \times \frac{\sigma}{eb}$. (a) $\alpha = 0.1b$ and (b) $\alpha = 0.2b$ [54].

The results depicted in **Figures 12** and **13** demonstrate that the increase in mobility becomes more pronounced with increasing charge carrier density, both for lower and higher densities, at the specified temperatures. However, at a temperature of $T = 235K$, the mobility becomes independent of density in a narrow range of low densities between 10^{14} and $4 \times 10^{14}\ cm^{-3}$. The reason for the increase in mobility with charge carrier density in a disordered system with Gaussian energy distribution is linked to the filling of deep states, as described earlier, which is influenced by the density's impact on the Fermi energy and the localization length. When the density increases, the deep states become occupied, causing an upward shift in the Fermi energy. Additionally, higher charge carrier densities can also increase the localization length [58]. These factors contribute to an increased average hopping rate of charge carriers, resulting in higher charge carrier mobility. Therefore, the variation of mobility with density is more significant when the disorder energy's energy distribution width σ is larger or when the temperature is lower. Our simulation results in the higher density region align with experimental findings [14, 15, 38]. However, the results obtained in the lower density region, although similar to the numerical results reported in [16], differ from the general conclusion drawn in [14, 15], which states that mobility is independent of density in the low-density region. The deviations observed in comparison with the experimental results can be attributed to the fact that the effects of electric field and density were not separated in the experiment. The influence of density on mobility, particularly in the low-density region, can be overshadowed by the effects of the electric field, which becomes evident when examining the curves for two different electric fields displayed in **Figures 12** and **13**. The dependence of charge carrier density on mobility is more prominent in **Figure 13**,

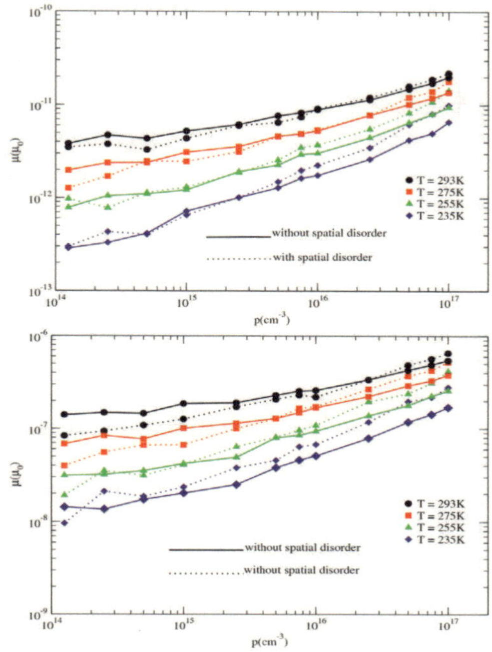

Figure 13.
Logarithm of the charge carrier mobility as a function of the charge carrier density for the same disorder energy, at different temperatures T and α, for the same electric field $F = 0.05 \times \frac{\sigma}{eb}$. *(a)* $\alpha = 0.1b$ *and (b)* $\alpha = 0.2b$ *[54].*

where the electric field is $F = 0.05 \times \frac{\sigma}{eb}$, compared to **Figure 12**, where $F = 0.1 \times \frac{\sigma}{eb}$. The electric field value used in these experiments was closer to the latter case, which results in a suppressed influence of charge carrier density.

The effect of density on mobility is enhanced in the presence of spatial disorder when comparing the dashed curves to their corresponding solid curves in both **Figures 12** and **13** at each temperature. Additionally, for a larger localization length ($\alpha = 0.2b$), we see that the dashed curves are positioned above the corresponding solid curves at low temperatures, while the opposite is observed at high temperatures. Moreover, in general, as the temperature decreases from high to low, each dashed curve tends to shift from the bottom to the top of the corresponding solid curve, similar to what is depicted in **Figures 10** and **11**. A similar trend can be observed in the kMC simulation results presented in Ref. [57] for mobility (with and without spatial disorder) versus electric field at $\hat{\sigma} = 2$, when the localization length increases from $0.1b$ to $0.5b$ for spatial disorder $\sigma_r = 0.01b$, while maintaining a low charge carrier density. The explanation for this behavior has already been discussed in relation to **Figures 10** and **11**.

4. Summary and conclusions

In this chapter, we have presented an analysis of the transport properties of charge carriers injected from metallic electrodes into systems characterized by both energetic and spatial disorders. To investigate the charge carrier mobility as a function of the external electric field, we focused on examining the effects of the spatial disorder parameter, which has traditionally been considered negligible and disregarded in previous studies [16, 23]. Additionally, we explored the influence of the localization length and other relevant parameters. To calculate the charge carrier mobility, we employed a variable range hopping model. This model was applied to both regularly spaced lattice sites and randomly distributed states, taking into account uncorrelated site energies following a Gaussian distribution. The randomly distributed states were generated using computer programs and uniformly distributed random numbers, allowing us to incorporate both spatial and orientational disorder effects. Through our simulations, we generated data on charge carrier mobility and investigated its dependence on temperature, applied electric field, and disorder parameters. Furthermore, we examined the effects of the spatial disorder parameter r and the ratio of localization length α to the lattice site parameter b on charge carrier mobility in systems with energetic and spatial disorder.

The results obtained for $\hat{\sigma} = 3$ and $r = 0.2$ demonstrate that there is no change in the mobility of charge carriers when the electric field varies. However, for $r = 0.25$, the mobility decreases with the electric field, which is consistent with previous studies [40, 41], but inconsistent with experimental findings. The disparity between our results and experimental values arises when the spatial disorder parameter r exceeds 0.15, and the localization length is $\alpha = 0.1b$. This indicates that the chosen value for $\alpha = 0.1b$ or $r > 0.15$, not only in our study but also in Refs. [40, 41], is not realistic from an experimental perspective. It is possible that either the considered localization length is smaller than the actual value, as proposed in previous studies [40, 41], or a spatial disorder parameter $r > 0.15$ is unreasonable when the localization length is $\alpha = 0.1b$. These findings imply that both the localization length and lattice parameter are important factors affecting the electric field dependence of charge carrier mobility in regular grids and spatially disordered lattice sites, as long as the spatial disorder parameter r is lower than or equal to the ratio of α/b. However, when r is higher relative to the ratio of α/b, the only parameter influencing the electric field dependence of charge carrier mobility is the localization length of disordered organic semiconducting materials. Furthermore, the mobility data generated in our study suggest that considering a localization length of $0.1b$ does not accurately reproduce experimental data. The agreement between our numerical results and experimental findings supports the assumption we made in modeling the device for simulation studies, particularly regarding the statistical nature of randomly distributed hopping sites. Based on this agreement, we can conclude that the material parameters describing the properties of charge carriers in the system, including the charge carrier localization length, energetic disorder, spatial disorder, and lattice site spacing, have been appropriately chosen.

Overall, this chapter provides insights into the transport properties of charge carriers in disordered systems, shedding light on the influence of spatial disorder, localization length, and other relevant parameters.

Acknowledgements

We would like to thank the International Programme in the Physical Sciences, Uppsala University, Uppsala, Sweden for the financial support to fulfill the necessary research facilities for this work. We are indebted to Mulugeta Bekele for their valuable discussions and comments to our work.

Conflict of interest

"The authors declare no conflict of interest."

Author details

Seyfan Kelil Shukri[1*] and Lemi Demeyu[2]

1 School of Applied Natural Science, Department of Applied Physics, Adama Science and Technology University, Adama, Ethiopia

2 Addis Ababa University, Addis Ababa, Ethiopia

*Address all correspondence to: seyfan74@gmail.com

References

[1] Burroughes JH, Bradely DDC, Brown AR, Friend RH, Burn PL, Holmes AB. Light-emitting diodes based on conjugated polymers. Nature. 1990;**347**: 539

[2] Sirringhaus H, Tessler N, Friend RH. Integrated optoelectronic devices based on conjugated polymers. Science. 1998; **280**:1741

[3] Sirringhaus H, Brown PJ, Friend RH, Nielsen NM, Bechgaard K, Langeveldvos BMW, et al. Two-dimensional charge transport in self-organized, high mobility conjugated polymers. Nature. 1999;**401**:685

[4] Dimitrakopoulos CD, Malenfant PRL. Organic thin film transistors for large area electronics. Advanced Materials. 2002;**14**:99

[5] Sariciftci NS, Braun D, Zhang C, Srdanov VI, Heeger AJ, Stucky G, et al. Semiconducting polymer-buckminsterfullerene heterojunctions: Diodes, photodiodes, and photovoltaic cells. Applied Physics Letters. 1993;**62**:585

[6] Bässler H. Charge transport in disordered organic photoconductors. A Monte Carlo simulation study. Physica Status Solidi B. 1993;**175**:15

[7] Coehoorn R, Pasveer WF, Bobbert PA, Michels MAJ. Charge-carrier concentration dependence of the hopping mobility in organic materials with Gaussian disorder. Physical Review B. 2005;**72**:155206

[8] Cottar J, Bobbert PA. Calculating charge-carrier mobilities in disordered semiconducting polymers: Mean field and beyond. Physical Review B. 2006;**74**: 115204

[9] Martin SJ, Kambili A, Walker AB. Temperature and field dependence of the mobility of highly ordered conjugated polymer films. Physical Review B. 2003;**67**:165214

[10] Roichman Y, Tessler N. Charge transport in conjugated polymers: The influence of charge concentration. Synthetic Metals. 2003;**135**:443

[11] Kohary K, Cordes H, Baranovski SD, Thomas P, Yamasaki S, Hensel F, et al. One-dimensional hopping transport in disordered organic solids. II. Monte Carlo simulations. Physical Review B. 2001;**63**:094202

[12] Baranovskii SD. Mott lecture: Description of charge transport in disordered organic semiconductors: Analytical theories and computer simulations. Physica Status Solidi A. 2018;**215**:1700676

[13] Gill WD. Drift mobilities in amorphous charge-transfer complexes of trinitrofluorenone and poly-n-vinylcarbazole. Journal of Applied Physics. 1972;**43**:5033

[14] Tanase C, Meijer EJ, Blom PWM, de Leeuw DM. Unification of the hole transport in polymeric field-effect transistors and light-emitting diodes. Physical Review Letters. 2003;**91**:216601

[15] Tanase C, Blom PWM, de Leeuw DM. Origin of the enhanced space-charge-limited current in poly (p-phenylene vinylene). Physical Review B. 2004;**70**:193202

[16] Pasveer WF, Cottaar J, Tanase C, Coehoorn R, Bobbert PA, Blom PWM, et al. Unified description of charge-carrier mobilities in disordered

semiconducting polymers. Physical Review Letters. 2005;**94**:206601

[17] Craciun NI, Wildeman J, Blom PWM. Universal Arrhenius temperature activated charge transport in diodes from disordered organic semiconductors. Physical Review Letters. 2008;**100**:056601

[18] Blom PWM, de Jong MJM, van Munster MG. Electric-field and temperature dependence of the hole mobility in poly (p-phenylene vinylene). Physical Review B. 1997;**55**:R656

[19] Servati P, Nathan A, Amaratunga GAJ. Generalized transport-band field-effect mobility in disordered organic and inorganic semiconductors. Physical Review B. 2006;**74**:245210

[20] Wang L, Fine D, Basu D, Dodabalapur A. Electric-field-dependent charge transport in organic thin-film transistors. Journal of Applied Physics. 2007;**101**:054515

[21] Partmeier L, Richert R, Bässler H. Poole-Frenkel behavior of charge transport in organic solids with off-diagonal disorder studied by Monte Carlo simulation. Synthetic Metals. 1990; **37**:217

[22] Novikov SV, Dunlap DH, Kenkre VM, Parris PE, Vannikov AV. Essential role of correlations in governing charge transport in disordered organic materials. Physical Review Letters. 1998; **81**:4472

[23] Zhou J, Zhou YC, Zhao JM, Wu CQ, Ding XM, Hou XY. Carrier density dependence of mobility in organic solids: A Monte Carlo simulation. Physical Review B. 2007;**75**:153201

[24] Demeyu L, Stafström S, Bekele M. Monte Carlo simulations of charge carrier mobility in semiconducting polymer field-effect transistors. Physical Review B. 2007;**76**:155202

[25] Mehraeen S, Coropceanu V, Brédas J-L. Role of band states and trap states in the electrical properties of organic semiconductors: Hopping versus mobility edge model. Physical Review B. 2013;**87**:195209

[26] Hulea IN, Brom HB, Houtepen AJ, Vanmaekelbergh D, Vanmaekelbergh D, Kelly JJ, et al. Wide energy-window view on the density of states and hole mobility in poly (p-phenylene vinylene). Physical Review Letters. 2004;**93**:166601

[27] Miller A, Abrahams E. Impurity conduction at low concentrations. Physics Review. 1960;**120**:745

[28] Verbeek G, Van der Auweraer M, De Schryver FC, Geelen C, Terrell D, De Meuter S. The decrease of the hole mobility in a molecularly doped polymer at high electric fields. Chemical Physics Letters. 1992;**188**(1-2):85-92

[29] Frenkel J. On pre-breakdown phenomena in insulators and electronic semi-conductors. Physics Review. 1938; **54**:647. [CrossRef]

[30] Martens HCF, Blom PWM, Schoo HFM. Comparative study of hole transport in poly (p-phenylene vinylene) derivatives. Physical Review B. 2000;**61**: 7489

[31] Mozer AJ, Sariciftci NS, Pivrikas A, Österbacka R, Juška G, Brassat L, et al. Charge carrier mobility in regioregular poly (3-hexylthiophene) probed by transient conductivity techniques: A comparative study. Physical Review B. 2005;**71**(3):035214

[32] Gartestein YN, Conwell EM. High-field hopping mobility in disordered

molecular solids: A Monte Carlo study of off-diagonal disorder effects. The Journal of Chemical Physics. 1994;**100**: 9175

[33] Rakhmanova SV, Conwell EM. Electric-field dependence of mobility in conjugated polymer films. Applied Physics Letters. 2000;**76**:3822

[34] Gartestein YN, Conwell EM. High-field hopping mobility in molecular systems with spatially correlated energetic disorder. Chemical Physics Letters. 1995;**245**:351

[35] Dunlap DH, Parris PE, Kenkre VM. Charge-dipole model for the universal field dependence of mobilities in molecularly doped polymers. Physical Review Letters. 1996;**77**:542

[36] Parris PE. Low-field hopping among randomly-distributed sites with uncorrelated energetic disorder. The Journal of Chemical Physics. 1998;**108**: 218

[37] Parris PE, Kenkre VM, Dunlap DH. Nature of charge carriers in disordered molecular solids: Are polarons compatible with observations? Physical Review Letters. 2001;**87**:126601

[38] Brondijk JJ, Maddalena F, Asadi K, van Leijen HJ, Heeney M, Blom PWM, et al. Carrier-density dependence of the hole mobility in doped and undoped regioregular poly (3-hexylthiophene). Physica Status Solidi B. 2012;**249**:138

[39] Vissenberg MCJM, Matters M. Theory of the field-effect mobility in amorphous organic transistors. Physical Review B. 1998;**57**:12964

[40] Nenasheev AV, Oelerich JO, Dvurechenskii AV, Gebhard F, Baranovskii SD. Fundamental characteristic length scale for the field dependence of hopping charge transport in disordered organic semiconductors. Physical Review B. 2017;**96**:035204

[41] Oelerich JO, Nenashev AV, Dvurechenskii AV, Gebhard F, Baranovskii SD. Field dependence of hopping mobility: Lattice models against spatial disorder. Physical Review B. 2017; **96**:195208

[42] Fichthorn KA, Weinberg WH. Theoretical foundations of dynamical Monte Carlo simulations. The Journal of Chemical Physics. 1991;**95**:1090

[43] Shukri SK. Charge transport properties in disordered organic semiconductors: Monte Carlo Simulation [doctoral dissertation] Addis Ababa University. 2021

[44] Baranovskii SD. Theoretical description of charge transport in disordered organic semiconductors. Physica Status Solidi B. 2014;**251**:487

[45] Hamadani BH, Richter CA, Gundlach DJ, Kline RJ, McCulloch I, Heeney M. Influence of source-drain electric field on mobility and charge transport in organic field-effect transistors. Journal of Applied Physics. 2007;**102**:044503

[46] Schönherr G, Bässler H, Silver M. Dispersive hopping transport via sites having a Gaussian distribution of energies. Philosophical Magazine B. 1981;**44**:47

[47] Züfle S, Christ N, Kettlitz SW, Valouch S, Lemmer U. Influence of temperature-dependent mobilities on the nanosecond response of organic solar cells and photodetectors. Applied Physics Letters. 2010;**97**(6):063306

[48] Yogev S, Halpern E, Matsubara R, Nakamura M, Rosenwaks Y. Direct

measurement of density of states in pentacene thin film transistors. Physical Review B. 2011;**84**(16):165124

[49] Melianas A, Kemerink M. Photogenerated charge transport in organic electronic materials: Experiments confirmed by simulations. Advanced Materials. 2019;**31**:1806004

[50] Shukri S, Deja L. Effects of localization length and spatial disorder on a charge carrier mobility in organic disordered semiconductors. AIP Advances. 2021;**11**:095118

[51] Schein LB, Peled A, Glatz D. The electric field dependence of the mobility in molecularly doped polymers. Journal of Applied Physics. 1989;**66**(2):686-692

[52] Hirao A, Nishizawa H, Sugiuchi M. Diffusion and drift of charge carriers in molecularly doped polymers. Physical Review Letters. 1995;**75**:1787

[53] Fumagalli L, Binda M, Natali D, Sampietro M, Salmoiraghi E, Di Gianvincenzo P. Dependence of the mobility on charge carrier density and electric field in poly (3-hexylthiophene) based thin film transistors: Effect of the molecular weight. Journal of Applied Physics. 2008;**104**(8):084513

[54] Shukri SK, Deja LD. Charge carriers density, temperature, and electric field dependence of the charge carrier mobility in disordered organic semiconductors in low density region. Condensed Matter. 2021;**6**:38

[55] Fishchuk II, Kadashchuk AK, Genoe J, Ullah M, Singh HSTB, Sariciftci NS, et al. Temperature dependence of the charge carrier mobility in disordered organic semiconductors at large carrier concentrations. Physical Review B. 2010;**81**:045202

[56] Arkhipov VI, Heremens P, Emelianova EV, Bässler H. Effect of doping on the density-of-states distribution and carrier hopping in disordered organic semiconductors. Physical Review B. 2005;**71**:045214

[57] Kaiser W, Albes T, Gagliardi A. Charge carrier mobility of disordered organic semiconductors with correlated energetic and spatial disorder. Physical Chemistry Chemical Physics. 2018;**20**:8897

[58] Gao W, Adinehloo D, Li X, Mojibpour A, Yomogida Y, Hirano A, et al. Band structure dependent electronic localization in macroscopic films of single-chirality single-wall carbon nanotubes. Carbon. 2021;**183**:774

Chapter 6

Epitaxial Growth of Semiconductor Alloys by Computational Modeling

Heber Vilchis, César Camas and Jorge Conde

Abstract

The research on semiconductor materials has attracted significant interest of researchers in the last few decades, due to their applications in numerous industrial sectors as well as in common life. Particularly, the investigation of these compounds has been improved with the support of the computational model theory. This chapter discusses the application of Molecular Dynamics, a classical computational approach, in the analysis of significant three-dimensional (3D) semiconductor material systems, emphasizing III-V and CIS (copper, indium or gallium as well as sulfur or selenium) families. Specifically, the growth processes and their thermodynamics phenomena are focused on the Molecular Beam Epitaxy (MBE) technique. In addition, this chapter provides considerations about the structural properties. We examine the evolution of the growth process and several important characteristics of the model that provides fundamental information on the structural defects.

Keywords: molecular dynamics, III-nitrides, CIS semiconductors, molecular beam epitaxy, cubic InGaN

1. Introduction

The development of commercially viable light-emitting diodes (LEDs), whose success was achieved in the 1990s by the analysis of III-nitride (III-N) compounds, marked a milestone of great importance in the family of III-V materials [1]. From those years until now, many of the advances in optoelectronics cannot be described without this group of materials. In particular, there is great interest in indium gallium nitride (InGaN), as it has several attributes that will make it an appropriate material for the fabrication of white solid-state devices, whose applications can be found in lighting, power electronics, and tandem solar cells, using it as heterostructures of $In_xGa_{1-x}N$/GaN (gallium nitride) [2, 3]. One of the fundamental characteristics of InGaN is its direct band gap (Eg), which can be adjusted from values of 0.65 eV (infrared) to 3.42 eV (ultraviolet), passing through the violet-blue, green, and yellow spectral region, depending on the molar fraction of indium (indicated by the subscript x in the stoichiometric formula) in compound [4]. Other important characteristics are the high mobility of the n-type carrier, resistance to radiation, and the optical absorption of the order of 10^5 cm^{-1} near the edge of the band [5, 6]. **Figure 1** shows the

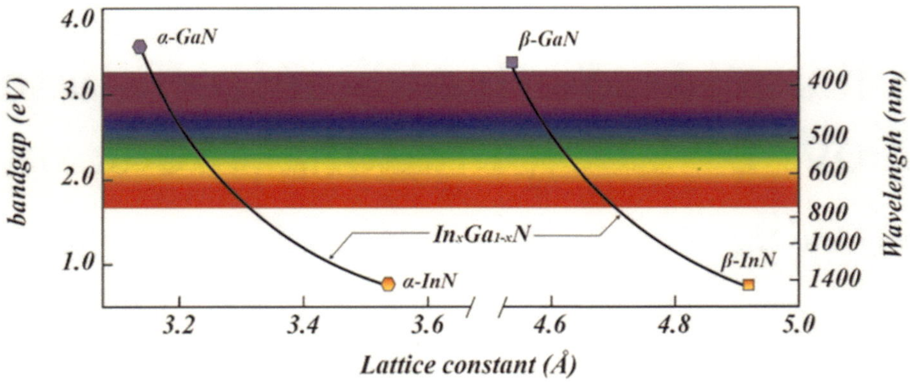

Figure 1.
Bandgap versus lattice constant of $In_xGa_{1-x}N$. The left and right lines correspond to the stable and metastable phases of the material, respectively.

bandgap versus the lattice constant of the compound, as well as the regions of the visible light spectrum, where solid lines represent the ternary values.

However, there is still a gap in the knowledge of some properties of $In_xGa_{1-x}N$, since they are calculated based on the binary compounds like gallium nitride and indium nitride (InN), such as the bandgap, the parameters of the lattice, elastic constants, effective electron mass, among others [7]. In addition, the state of the art of the material has focused mainly on the study of the hexagonal structure, because it is the stable thermodynamic phase. However, it is estimated that better properties could be obtained if the material grows in a cubic phase (metastable). This suggests some technological challenges that make it difficult for the effective use of $In_xGa_{1-x}N$ alloys for tandem photovoltaics and other applications.

In addition, one of the central drawbacks in $In_xGa_{1-x}N$ research is the critical reduction in quantum efficiency at alloy compositions greater than x $\sim$ 0.3. This problem is attributed to multiple underlying mechanisms, such as the high density of dots and linear defects as a result of interlayer lattice mismatch and the low growth temperatures required to achieve high indium (In) content [8]. Another mechanism is the spinodal decomposition in the intermediate In content that results in films with mixed hexagonal and cubic phases [8]. Furthermore, a fundamental mechanism given in the hexagonal phase is the spontaneous and piezoelectric polarization that induces electric fields in the opposite direction to the depletion field of a diode [9]. In contrast, the cubic phase (zincblende) does not exhibit polarization due to its centrosymmetric structure [10]. Also, as a result of the reduced phonon scattering rate due to a higher degree of crystal symmetry, it is intuited that cubic nitrides could have a higher saturated electron drift [11].

Moreover, the most common methods to deposit $In_xGa_{1-x}N$ are Metal–Organic Chemical Vapor Deposition (MOCVD) [12, 13] and Molecular Beam Epitaxy (MBE) [14, 15]; although Plasma-Assisted Atomic Layer Deposition (PA-ALD) has also been used as an alternative to the growth technique, since the formation of nanoscale thin films is well controlled on a substrate and this allows a specific thickness at atomic dimensions, as a result of self-limited surface reactions [16, 17]. However, there are some challenges in its application in nitride semiconductors and specifically little research on $In_xGa_{1-x}N$ supporting film growth phenomena by this technique [18].

Hence, it is clear that the understanding of the relationship between the characteristics of $In_xGa_{1-x}N$ and the performance of the device is useful to obtain ternary heterostructures with high efficiency, preferably in the cubic phase. Theoretically, developing devices with such technology, specifically solar cells, would represent efficiencies that could reach values lose to 70%, considering multilayer or tandem structures [19], although the Shockley-Queisser (SQ) limit for homojunction structures is estimated at 33.77% [20]. Until now, the technology based on $In_xGa_{1-x}N$ with pure cubic phase has not been achieved.

Due to how complex and costly the experimental growth process of $In_xGa_{1-x}N$ can be, a great alternative to study the behavior during its deposition process is the application of theoretical-computational simulations.

In this chapter, we present a detailed description of the epitaxial growth in semiconductor alloys through computational modeling. Our exploration begins by delving into Molecular Dynamics Simulation, an impactful tool that allows us to unravel the intricate atomic-scale processes involved in epitaxial growth. At that point, our focus centers on epitaxial growth simulation models and their analysis of $In_xGa_{1-x}N$ systems, providing a comprehensive understanding of the complex dynamics driving alloy crystal formation. Nonetheless, we recognize the limitations of these growth models while highlighting their constraints and areas for improvement. Subsequently, the chapter transitions to explore the significant influence of molar concentration of Indium on alloy performance, serving as a critical element. Additionally, we carry out a comprehensive study on dislocation analysis, revealing the influence of crystal defects on the quality of epitaxial growth. Finally, our exploration takes us to the domain of Cu-Se interactions during the growth process of copper-indium-gallium-diselenide compounds (CIGS), researching into the interrelationship between these constituents. Through this comprehensive exploration, we aim to provide valuable insights into the computational modeling of epitaxial growth for semiconductor alloys, offering a holistic understanding of this intricate process.

2. Molecular dynamics simulation

A wide variety of simulation techniques and methods have been developed over the years for the simulation of atomic and molecular systems, standing out for their wide utility of the Density Functional Theory (DFT), the Monte Carlo method (MC), and Molecular Dynamics (MD). Focus on MD, this is a deterministic method that allows the behavior or evolution of a system to be analyzed over time, calculating the forces between the atoms of the system using Newton's equations of motion; that is, it generates information about the position and velocity of each particle at each instant of time [21]. The general mathematical sequence for solving the positions and velocities at a given instant is described as

$$\text{Energy} \rightarrow \text{Force} \rightarrow \text{Aceleration} \rightarrow \text{Velocity} \rightarrow \text{Position}$$

$$\vec{F} = -\frac{dU}{dr}; \tag{1}$$

$$\vec{F} = m_i \vec{a}_i; \tag{2}$$

$$\vec{a}_i = \frac{d\vec{v}_i}{dt}; \tag{3}$$

$$\vec{v}_i = \frac{d\vec{r}_i}{dt} \tag{4}$$

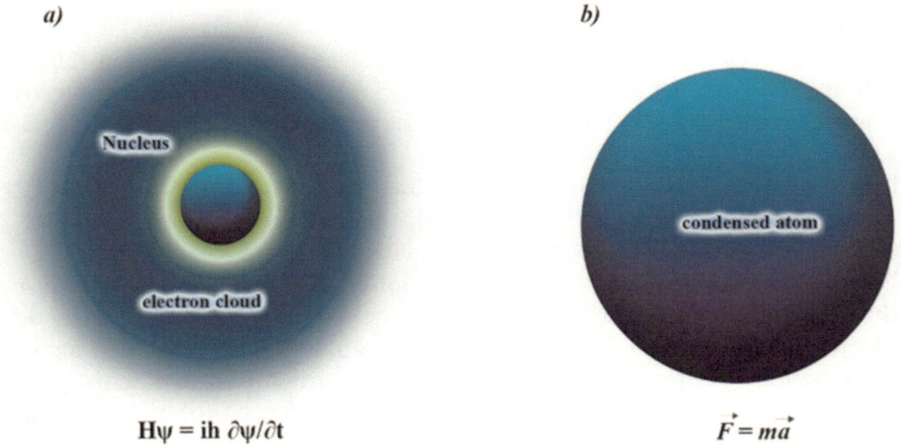

Figure 2.
The atomic model considered for a) density functional theory and b) molecular dynamics method.

where F is the sum of all the forces acting on an atom, dU is the atomic potential gradient, m_i, a_i, v_i, and r_i are the atomic mass, acceleration, velocity, and position of the ith atom, respectively. Therefore, knowing the potential of a system as a function of the interatomic distance, it is possible to calculate the strength of each atom within it and obtain the time evolution of the system under this series of algorithms.

The idea of using Molecular Dynamics comes from modeling the atom as a material point (an object completely described by classical mechanics), simplifying the calculations for a number of the order of 10^6 atoms or higher. This is possible because the role of the electrons can be neglected, so it is assumed that the electronic wave function instantly adapts to the current configuration of the atoms. However, it is noted that due to this simplification, obtaining information on the potential between atoms (interatomic potential) should be generated empirically [22]. **Figure 2** shows a comparison of the use of the atomic model in DFT, where the electron is crucial for numerical calculations, and the condensed atomic model of MD, which assumes the atom as a whole.

Likewise, the use of the classical approach is capable of obtaining many of the characteristics of macroscopic systems, such as structural and mechanical properties, defects and dislocations, surface diffusion, grain boundary structures, thermal capacity, and thermal conductivity, among others. In addition, it is possible to describe epitaxial deposition and growth processes by Sputtering, MOCVD, MBE, as well as nanoindentation. An initial configuration of the system is required to be analyzed; i.e., the initial positions of the atoms. Once the trajectories of these particles are known, the properties of the system can be computed as an average of statistical calculations.

As we can see, MD is a fast, reliable method that can analyze systems with a greater number of particles, if we compare it with those methods based on first principles. However, some limitations need to be taken into consideration when choosing this method, such as the availability of potentials of some systems of interest, the scale of time, and the size is still limited.

2.1 Interatomic potential: Stillinger-Weber

On one hand, since the calculations of the velocity and position of the atoms depend on the potential that exists between them, this interaction energy is described according to a mathematical function that contains fundamental parameters—mainly the distance between atoms and bond energy, according to the nature of the structures of these atomic systems, derived in conventional representations. This interaction between atoms is nothing more than a balance between the phenomena of repulsion and attraction, resulting in a state of minimum potential at a distance called equilibrium.

On the other hand, it is important to note that such mathematical descriptions come from both experimental data and first-principles calculations. The first principles implies that, although for the resolution of the behavior of atoms over time there are no quantum mechanical implications, information about the electronic cloud is implicitly found in the description of the potential from the calculation of the total force of the system. Potential equations can be grouped into interatomic and intermolecular, such as pair-pair, force field, multibody, and ionic solid interactions. The use of one potential or another, even more than one (hybridization), depends on the purposes of the process to be simulated, as well as the structure of the elements or compounds involved in it.

For the descriptive purposes of the atomic interactions involved in the simulations of III-nitride systems, the Stillinger-Weber (SW) potential has been widely used. This potential was conceived by Frank Stillinger and Thomas Weber for the recreation of semiconductor materials with tetrahedral bonds (diamond and zincblende structures) [23]. Originally developed for silicon, the potential overview has been adapted for other material systems, such as cadmium telluride (CdTe), molybdenum disulfide (MoS_2), and GaN and InGaN. The mathematical definition of the potential is given by.

$$E = \frac{1}{2}\sum_{i=1}^{N}\sum_{j=i_1}^{i_N}\left[V_{IJ}^{R}(r_{ij}) - V_{IJ}^{A}(r_{ij}) + u_{IJ}(r_{ij}) \sum_{\substack{k=i_1 \\ k\neq j}}^{i_N} u_{IK}(r_{ik})\left(\cos\theta_{jik} + \frac{1}{3}\right)^2\right] \quad (5)$$

Where i_1, i_2, ... , i_N are the surrounding atoms of the ith atom, θ_{jik} is the angle produced by the bonding effect of the i atom with j and k atoms. Also, the representation of repulsive and attractive energies is given by $V^{R}{}_{IJ}(r_{ij})$ and $V^{A}{}_{IJ}(r_{ij})$, respectively. $U_{IJ}(r_{ij})$ is the supplementary expression and suffixes ij and IJ designate a pair of atoms and their species, correspondingly. These functions are described by seven dependent pair parameters (ε, σ, a, λ, γ, A, and B), and by two additional parameters p and q that generally have values of 4 and 0, respectively. It is important to note that the value of the product $a \cdot \sigma$ represents the cutoff distance because the pair functions mentioned above are suppressed when $r = a \cdot \sigma$.

3. Epitaxial growth simulation models of $In_xGa_{1-x}N/GaN$

3.1 Simulation conditions

To identify the phenomena involved in the $In_xGa_{1-x}N$ deposition process, our MD simulations were performed using the Large-scale Atomic/Molecular Massively Parallel Simulator (LAMMPS) code. Likewise, the visual processing and analysis of the results were carried out with the open-source program Open Visualization Tool

(OVITO), using the “diamond structure identifier” modifier for the analysis of the cubic phase percentage; the “Dislocation Extraction Algorithm (DXA)" modifier to identify and calculate the length of dislocations in different molar fractions, as well as the “Surface Mesh Construction” modifier for calculating the roughness of the models.

3.2 Interatomic potential for $In_xGa_{1-x}N$

We use the Stillinger-Weber In-Ga-N potential of Zhou and Jones [24, 25] for the development of the ternary systems, to simulate the relations of the atoms implicated in the development of the Molecular Dynamics. The potential implies a penalization for nontetrahedral bond angles, providing minimum energy for tetrahedral structure types (diamond, wurtzite, zincblende). Therefore, the mathematical expressions of this potential calculate with stability the $In_xGa_{1-x}N$ crystal growth process. **Table 1** summarizes the fundamental SW potential parameters for In-Ga-N interactions.

3.3 Buffer layer considerations

It is well known that a GaN buffer layer is deposited before the epitaxial growth process of cubic $In_xGa_{1-x}N$, to reduce the stress between the ternary films and the substrate, due to the high percentage of lattice mismatch that exists between them, because that induces hexagonal phase defects and inclusions. For this reason, a square-based GaN zincblende-type structure with dimensions of 103.96 Å (x,y) had been created; while, for the z growth direction, the dimension was 22.6 Å. This undoped GaN (c-GaN) buffer layer consisted of 22,218 particles (11,109 gallium (Ga) atoms and 11,109 nitrogen (N) atoms). To correctly configure the simulation characteristics, the buffer layer had been divided into three regions. **Figure 3** shows the general growth pattern of InGaN in a buffer layer of c-GaN. The blue spheres represent the fixed region, the red spheres indicate the thermal region, and the white spheres represent the free or Newtonian atoms. The orange, yellow, and purple spheres represent indium, gallium, and nitrogen atoms, respectively.

Each of the buffer layer’s regions is described below:

1. Fixed or immobile layer. By depositing adatoms on the surface of the substrate with a certain velocity (incidence energy), they will tend to transfer momentum and will move the surface atoms of the substrate, causing possible ruptures in its

Parameter	In-N	Ga-N	In-Ga	In-In	Ga-Ga	N-N
σ	1.8527	1.7159	1.7691	1.9383	1.7596	1.7269
ε	2.2020	2.2896	1.9843	2.4498	2.9263	4.4201
γ	1.2000	1.2000	1.2000	1.2000	1.20000	1.2000
λ	32.5000	32.5000	32.5000	32.5000	32.5000	32.5000
A	7.0496	7.0496	7.0496	7.9170	7.9170	7.0496
B	0.7615	0.6410	0.8659	0.9700	0.9700	0.9698
a	1.7999	1.7996	1.7109	1.6222	1.6071	1.6300

Table 1.
Fundamental parameters of SW potential for In-Ga-N interactions.

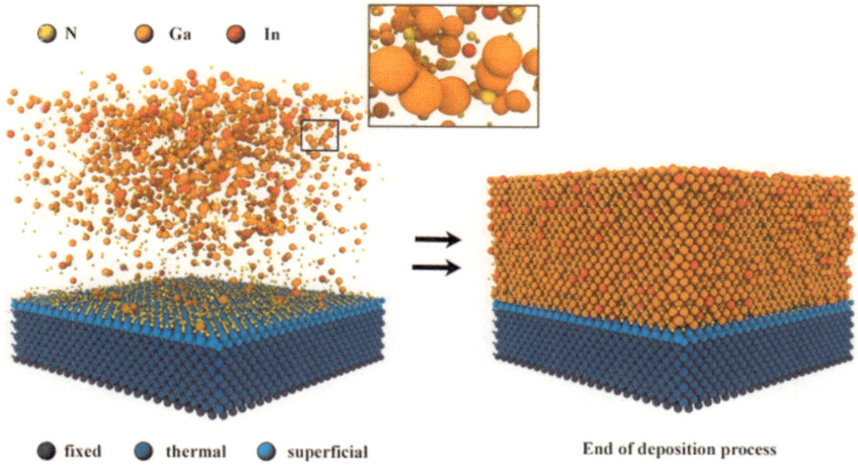

Figure 3.
Gallium nitride (GaN) buffer layer and its corresponding regions.

structure. For this reason, during the simulation, the atoms in the lower layer of the substrate are fixed to prevent this phenomenon. Therefore, a region consisting of two monolayers (Ga and N in x-y planes) was created in the lower part to avoid this effect in the simulation.

2. An intermediate region is also called thermal bath. It represents the thermal control through a canonical ensemble (NVT) for the development of the integration time applying a Nosé-Hoover thermostat. For all cases considered, the substrate temperature was set at 550°C, which is typical for MBE experimental processes.

3. Free region or surface interaction. The last region or the surface region also consisted of two Ga-N monolayers, where the interaction naturally occurs between the surface atoms of the buffer layer and the adatoms that reach them (solid–gas interface), establishing bonds of growth that will produce the epitaxial structure. A microcanonical ensemble (NVE) governs this region.

3.4 Border conditions

On one hand, periodic boundary conditions were established for the x and y directions, as well as a fixed volume to impose the bulk lattice constant of the substrate based on the assumption that the GaN buffer layer is much thicker than the film, such that the lateral dimensions of the system are determined by the buffer layer and thus will define the lattice mismatch between the $In_xGa_{1-x}N$ film and the GaN buffer layer. On the other hand, for the z-direction, a nonperiodic boundary condition (free) was established to carry out the epitaxial growth.

3.5 Velocity system conditions

The N, Ga, and In atoms were ejected from a specific region near the top of the simulation cell to the GaN (001) interlayer in different deposition ratios depending on the molar fraction of In ($0 \leq x < 0.5$). The In, Ga, and N atoms have velocities of 4.12, 5.28, and 11.8 Å/ps, respectively. These velocities correspond to the incident energy of 0.1 eV at a sublimation temperature of 900°C according to the mass of each element. The atoms were deposited one group after another in this order: first, a group of 1500–2500 nitrogen atoms, emulating a nitridation process; second, the ejection of clusters of indium, gallium, and nitrogen atoms at the specified timesteps, so the deposition ratio was directly proportional to the In molar percentage. In laboratory conditions, the deposition ratio was lower than the value used for this simulation. Nonetheless, the computational limitations require a reasonable time of execution and analysis, thus, this change in deposition ratio is logical and indispensable. Moreover, an extra thermal subprocess (annealing) was applied to ensure that the incident atoms were diffused properly on the buffer layer surface. The last N, Ga, and In atoms were ejected at different timesteps, due to indium molar fraction. However, its deposition is not instantaneous on the buffer layer. Therefore, the total timesteps of each model were extended from $50x10^3$ to $250x10^3$ timesteps (as more In atoms are deposited, more additional timesteps are required) to make sure that the total number of atoms was deposited.

4. Growth model limitations

Although significant theoretical-computational results were obtained, the simulations of the $In_xGa_{1-x}N$ growth models on c-GaN buffer layers present some limitations related to the time needed to make sense of them; this means that the addition of timesteps was considered after the time required to eject all the atoms toward the buffer layer, to guarantee the deposit of the adatoms on the surface. Therefore, more computational time is needed to complete the process. The time adaptation for each simulation model depends on the molar fraction of indium. That is, the higher the molar fraction of indium was declared in the simulation, the higher volume of timesteps is required. In addition, the relaxation stage of the grown structures represented an extra time of $100x10^3$ to $200x10^3$ timesteps, since it is necessary to bring them to room temperature conditions at the end of the deposition process. Without this time range, it would mean the modification of the structure or the appearance of new dislocations in the growth volume. Experimentally, the growth times are also adjusted according to the concentration of indium; the growth rate tends to be decreased to obtain layers of better crystalline quality as x tends to 1. In this sense, the growth rate of the simulations had a decreasing tendency from $x = 0$ to $x = 1$ and, therefore, the simulation time was extended for each model as a guarantee of obtaining the material layers.

5. Indium molar concentration effect

The indium concentration (molar fraction x) is critical to obtaining $In_xGa_{1-x}N$ thin films with a good quality structure. In this investigation, several samples (simulation models) of $In_xGa_{1-x}N$ with arbitrary In concentrations were produced, by the zero-to-one experimental values used by Vilchis et al. [11] and V.D. Compeán et al. [26]. In this sense, we recognize three groups of computational models of interest that will be analyzed in detail below. It should be noted that the models were achieved by the feasibility provided through the SW potential, which recreated the presence of strong tetrahedral angular preferences for this type of covalent compound.

5.1 $0 \leq x \leq 0.29$ models

This group of molar fractions of indium represents the models in which, theoretically and experimentally, the preferential value of the cubic structure has been analyzed, using suitable conditions of low temperature (550°C) and a buffer layer consistent with similar crystallography. However, the appearance of a low percentage of structures like wurtzite in the first models is imminent. **Figure 4** shows the epitaxial models with a molar percentage x between 0 and 0.21; each In concentration indicates the cubic and hexagonal $In_xGa_{1-x}N$ phases.

As shown in **Figure 4**, for the binary GaN growth model, the cubic structure is present throughout the volume of the film. This is possible thanks to the fact that the bond potentials favor the formation of the zincblende tetrahedral structure, so that the accommodation of the atoms (diffusion process) at the sites on the surface of the c-GaN buffer layer has been faster and adequately conditioned with no significant changes in their positions. Therefore, the extraordinary settling and relaxation times have been considerably short for this particular model ($50x10^3$ timesteps). On the other hand, from the ejection of indium atoms to form the first ternary alloys, the values where x were 0.11, 0.17, and 0.21 tended to show features in the superficial part

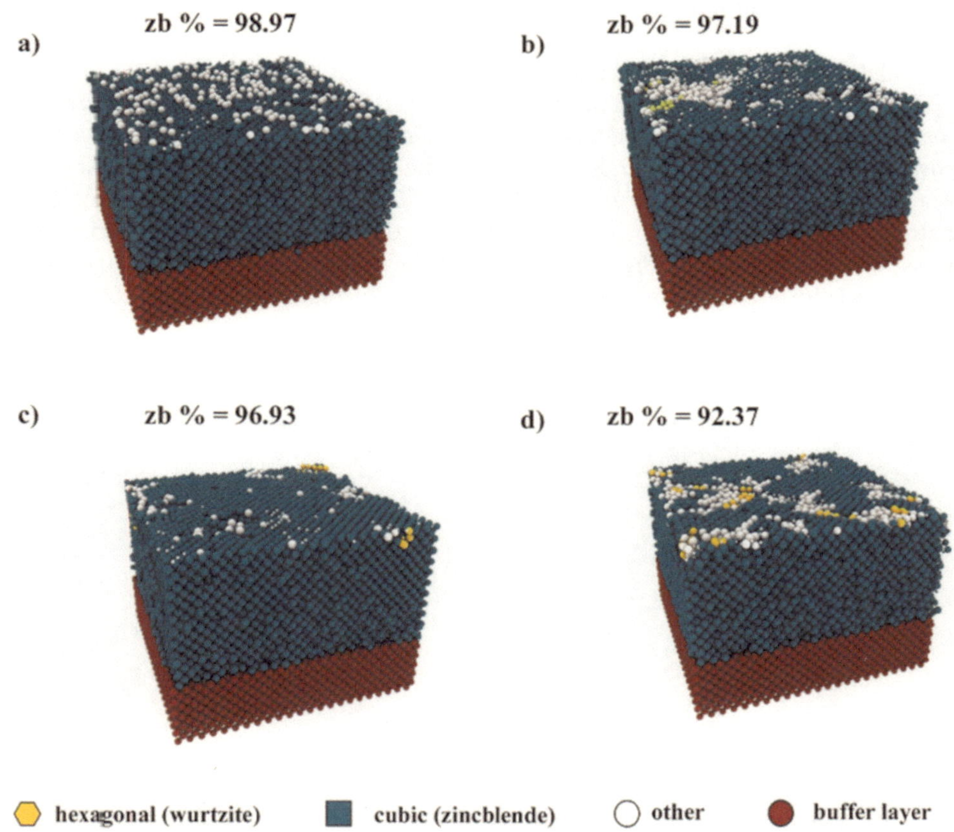

Figure 4.
Computational models of $In_xGa_{1-x}N$/GaN (100) with In molar fractions of: a) 0, b) 0.11, c) 0.17, and d) 0.21.

of hexagonal embedding, in percentages equivalent to 0.34, 1.41, and 1.44%, respectively. This is due to the phenomenon of indium segregation; atoms of this nature tend to separate toward the surface, changing the stacking sequences in the last grown layers.

Similarly, concerning the models with values where $0.21 < x \leq 0.29$, these show a drastic change from values higher than 25%, since under this fraction the trend of wurtzite incrustation remains at low levels ($\sim$ 1.5% presence). However, as the values tend to 30%, as in the recreated model of 29%, the presence of the hexagonal phase tripled, making its presence more noticeable in the grown volume. **Figure 5** shows the models of these two-mole fractions of indium.

In general, an increment in the value of indium atoms in the models implies a decrease in the cubic structure percentage during the growth process. A more specific observation of the $In_xGa_{1-x}N$ models with lower values than 21% shows that their structures have a preferential zincblende type of stacking sequence, in concordance to V.D. Compeán et al. [26]. The inclusion of wurtzite was minimal (5%). Likewise, it was observed that the inclusion of indium in the growing volume generates a greater number of dislocation lines, which induce a change in the stacking sequence in the atoms; this is reflected in an already experimentally identified rotation of 60° in the structure. Therefore, metastable growth conditions tend to break down, to establish the thermodynamic equilibrium of the compound.

5.2 $0.30 < x < 0.50$ models

Models with values greater than 30% were recreated. As the state of the art of $In_xGa_{1-x}N$ points out, there is a marked tendency to form the wurtzite structure from these intermediate or higher values of x. The recreated models of 32, 35, and 40% show that values are similar to those found in the models of 29%. This is most likely because the flux ratio of indium is practically indistinct compared to the total volume of atoms that are ejected across the surface. Since the thermodynamic conditions were not significantly modified, the arrangement of the indium atoms tends to maintain similar trajectories. Therefore, the hexagonal phase inclusion values are 4.66, 4.68, and 4.71%, respectively (**Figure 6**).

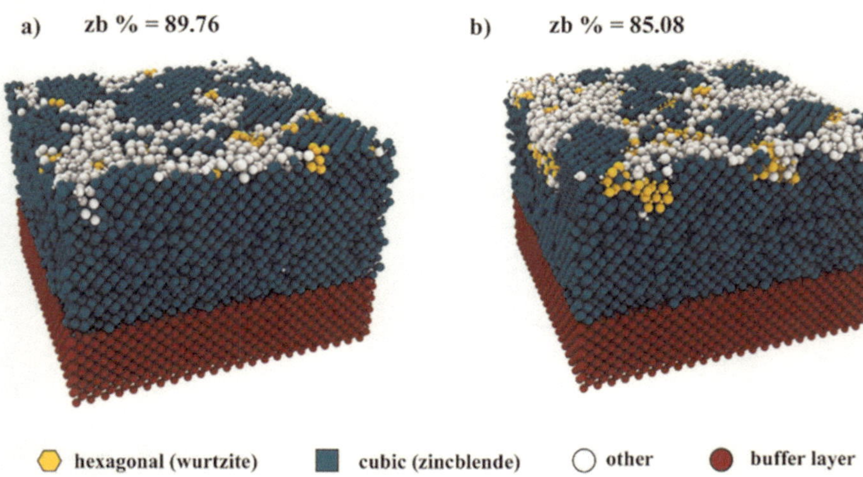

Figure 5.
$In_xGa_{1-x}N/GaN$ (100) models for In molar fractions of a) 25 and b) 29%.

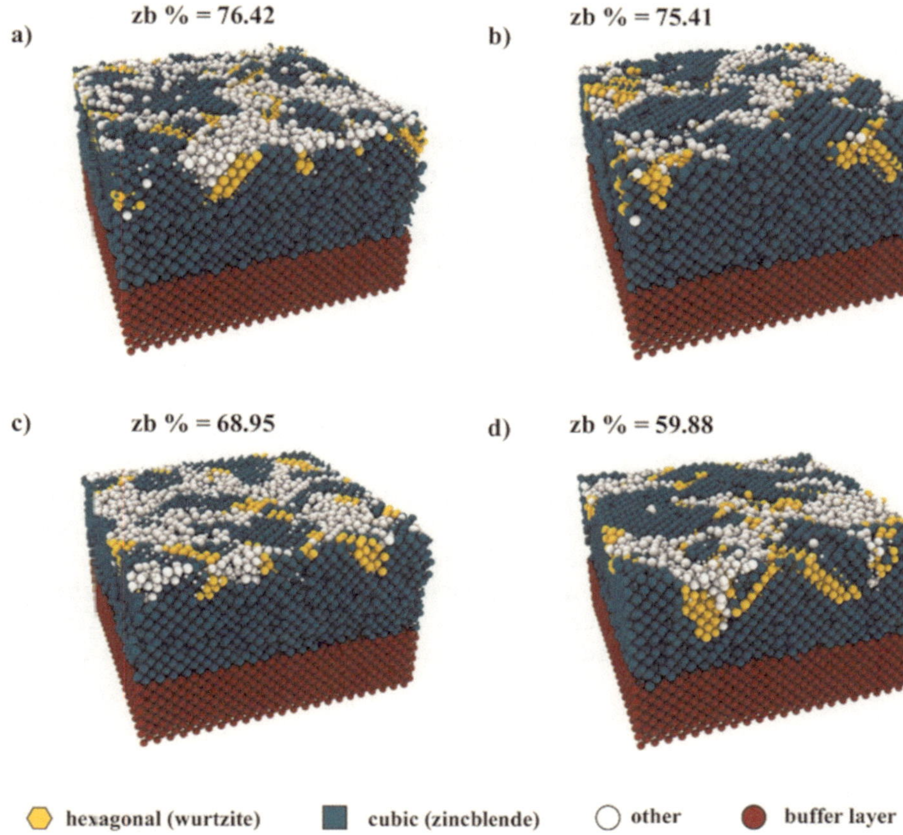

Figure 6.
$In_xGa_{1-x}N/GaN$ (100) models for In molar fractions of a) 32%, b) 35%, c) 40%, and d) 48%.

An increase in the molar fraction of indium, with intermediate values of x (48%), shows a visible change in the composition, since the tendency of a hexagonal nature increases by more than 3%. It is notable that, for this model, the flux ratio of indium is practically the same as that of gallium atoms. Thus, despite the velocity and the extraordinary time that is impregnated in the model ($200x10^3$ extra timesteps), they are not enough to avoid the preference in the In-N bonds of the Ga-N, modifying the sequence of stacking in the <111> directions, where the wurtzite structure tends to be preferred.

Furthermore, we found that between 29 and 48% In concentration, the deposited $In_xGa_{1-x}N$ structure drastically changed its cubic percentage, and an amount of wurtzite phase increased dramatically in the volume of the material [27]. This occurs due to the effect of indium flux since by increasing the number of atoms ejected from the adatoms region toward the growth surface, the temperature of the system also tends to increase due to the energy acquired in the process, which causes a segregation effect, changing the stacking sequences in some regions of the volume.

6. Dislocation analysis

As mentioned above, in experimental investigations, a GaN buffer layer has been used as a buffer film to reduce lattice mismatch and surface tension related to dislocation density. Therefore, we consider two GaN monolayers as a boundary to study the behavior of the hexagonal inclusion and its induced defect. We note an increase in dislocation density as the In concentration increases. So, we consider analyzing the dislocation with the "DXA modifier" tool of OVITO. The dislocation density (in units cm^{-2}) was calculated for each sample, taking into account the total length of the dislocation lines within the mass divided by the volume of material grown from $In_xGa_{1-x}N$ [28]. **Figure** 7 represents the models of the meshes and dislocation lines of molar fractions of indium in the range of $0 \leq x \leq 0.48$.

For the simulated volume of $In_xGa_{1-x}N$, only the time until the deposited atoms acquired a structure recognized by the "Diamond Structure Identification" modifier was considered. Therefore, the volumes of each calculation were directly associated with the critical thickness of the material. This led to the observation that an increase in the molar fraction of indium in the models induced a decrease in the critical thickness of the material. We observe that at an indium molar fraction lower than 0.29, the growth process tends to generate a low dislocation density and few vacancies in the atomic system, which make possible a significant percentage of zincblende structure. However, in proportions between 32 and 48%, the dislocation density tends to increase dramatically. Otherwise, a high tendency of Frank and Shockley partial defects results in the models with molar fractions up to 30%. This is an indicator of variations in the stacking sequence of III and V atom types, creating several faults proportional to the deposition ratio and the incremental presence of indium atoms.

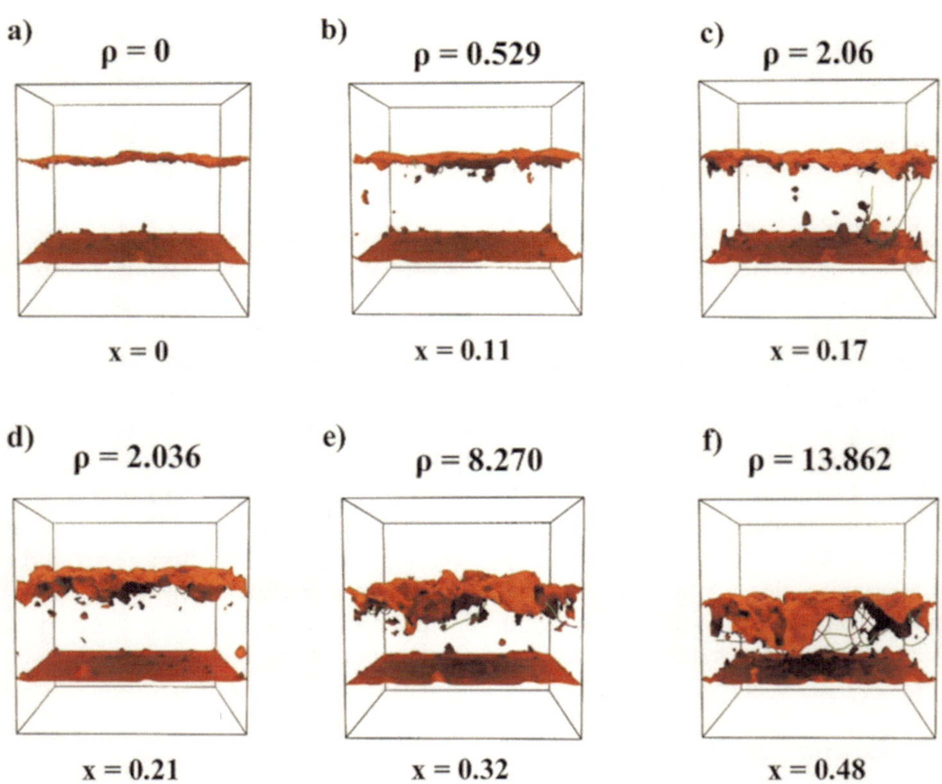

Figure 7.
Dislocation trend of $In_xGa_{1-x}N$ models for $0 \leq x \leq 0.48$ (dislocation density ρ in $cm^{-2} \times 10^{12}$).

These defects generate edge and interstice dislocations in these stacking regions, predominantly in the [111], [112], and [110] directions.

7. Cu-Se interactions for the CIGS growth process

Nowadays, there is a lack of a computational potential framework for copper-indium-gallium-diselenide compounds (CIGS); nevertheless, there exist theoretical approximations [29, 30]. From this point, we determined to analyze the copper-selenium interactions from a co-evaporation second-stage process using a hybrid/overlay LAMMPS scheme to express the Cu-Cu, Se-Se, and Cu-Se interactions. These energy and distance relations between copper and selenium were obtained from different kinds of potentials: Cu modified-embedded-atom [31], Cd-Te-Zn-Se-Hg-O Stillinger-Weber [32], and Morse [33]. The last was used for those interactions between different natures of atoms. These potentials assure the free exchanges of energy between Cu and Se atoms.

The simulation geometry comprises a square base of 54.1 Å Cu buffer layer, located in the x-y plane, with a 10.8 Å thickness (z-direction); this represents seven layers of 3150 atoms, as we can see in **Figure 8**. To control the conditions of the process, we divided the buffer layer into three principal groups: A) fixed group, consisting of two layers at the bottom to prevent the distortion of the structure due to the incident energy of the selenium atoms; B) thermal control group which means the implementation of a thermostat governed by the equations of the canonical NVT ensemble [34]. Therefore, the temperature was controlled between 325 K and 550 K. C) A superficial assembly, composed of two layers at the substrate's uppermost region, can freely interact and exchange energy with the atoms that are being ejected onto the substrate.

Figure 9 presents the superficial interactions of the copper and selenium atoms, observing the bonds between them. The Se atom's velocity was calculated at 3.28 Å/ps for a negative z-direction, according to the range of temperatures used for controlling the thermal conditions.

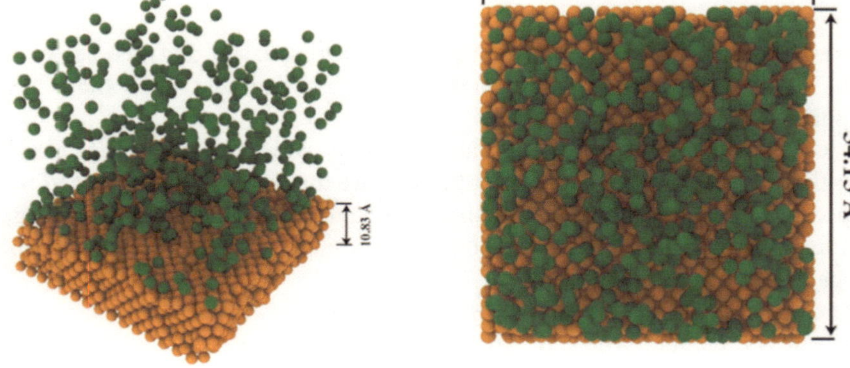

Figure 8.
a) Deposition process of Se (green) on top of the Cu (pink) buffer layer and b) top view.

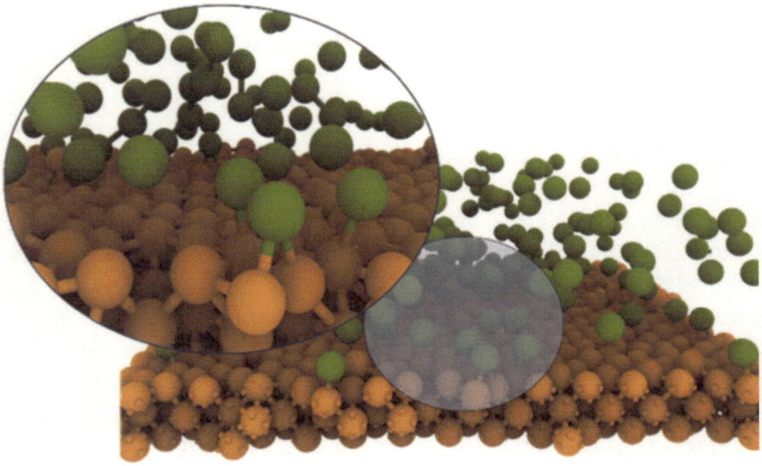

Figure 9.
Copper-selenium (Cu-Se) interactions at the surface. Velocity at 550 K, cutoff: 2.5 Å, 20k timesteps. Green and pink spheres represent selenium and copper atoms, respectively.

7.1 Morse interatomic potential

As we anticipated, the energy relation of Cu and Se atoms is governed by the Morse potential, which is a potential model of an atom/molecule particle and presents a similar performance to the Lennard Jones (LJ) potential: both of these energy potentials have a pair of repulsion and attraction functions; adjustable parameters and they are used in packed crystal structures [35].

Nevertheless, the Morse potential has better precision for diatomic molecule cases. Moreover, it is widely used in Molecular Dynamics simulations of some important metal materials such as copper [36, 37] or in the modeling of covalent structures. Additionally, the selection of this potential type corresponds to its use for diatomic molecules and interactions in gas–solid interfaces.

We can observe the description of the Morse potential in the following expression:

$$E = D_0\left[e^{-2\alpha(r-r_0)} - 2e^{-\alpha(r-r_0)}\right]; r < r_c \tag{6}$$

where D_0 means the separation energy, r is the internuclear distance between atoms, r_0 is the equilibrium bond distance, and α is a controlling factor that modifies the potential "width."

The cutoff radius r_c was considered an important parameter to study to clarify whether the cutting distance is required to estimate the total energy provided by Morse potential. Likewise, the cutting distances could signify an important influence of the interfacial properties, since the boundary conditions limit the potential energy in a range of $r < r_c$ [38].

The expectations of the models pointed out to cutoff radius close to or equal to 2.5 Å since it is considered the standard for LJ interactions. We established the model conditions around these values to obtain different simulations varying the cutoff radius (for example, we made models around 0.3 Å), observing in the visualization process the effect that these produced in the simulations. The results indicate that, at a smaller cutoff radius, in the range of 1 to 2.2 Å, the selenium atoms come into contact with deeper layers in the copper substrate, which distorts the copper buffer layer structure [39].

As well, according to the thermal control, we varied the selenium atom velocities. It was noticed that better results were reached when the temperature was set at 550 K and the cutoff was around 2.5 Å. These results are consistent with other theoretical and experimental Cu-Se interaction studies.

8. Conclusions

Molecular Dynamics method has demonstrated the advantages of its classical perspective. We can observe the trajectory of a group of atoms or molecules (atomic/molecular system) and predict their behavior through time; this means that some important characteristics, such as mechanical and structural, could be obtained with an appropriate thermodynamic scheme and adequate interatomic potential energy. Moreover, Molecular Dynamics replicate with significant reliability the phenomena of deposition material process. Specifically, the reproducibility of the MBE deposition technique has been possible, under typical thermodynamic conditions (550°C), in the growth of $In_xGa_{1-x}N$ on a GaN buffer layer previously structured in cubic phase for molar fractions of indium between $0 \leq x < 0.50$. The creation of the buffer layer with a previously GaN declared cubic structure as possible and the temperature control was through a canonical NVT assembly with a Nosé-Hoover thermostat. Likewise, the interatomic potential plays a transcendental role, since all the possible interactions that arise in the deposition process of this material have already been described based on the SW equations. Additionally, other semiconductor systems could be reproduced by a similar sequence of model simulations, as we can notice in Cu-Se interaction analysis for CIGS structures. Nevertheless, it is important to declare that Molecular Dynamics has some limitations for the study of some materials like semiconductors, due to the lack of potential for some particular energy interactions, the need for more computational processing capacity for greater systems (grain border conditions), and thus, the time to achieve reliable results.

Acknowledgements

This work was partially supported by Conacyt – Frontier Science 2019 project # 1046631.

Conflict of interest

We declare that we have no known competing financial interests or personal relationships that could have influenced the work reported in this book chapter.

Author details

Heber Vilchis[1]*, César Camas[2] and Jorge Conde[3]

1 Instituto de Investigación e Innovación en Energías Renovables, Universidad de Ciencias y Artes de Chiapas, Tuxtla Gutiérrez, Chiapas, México

2 Ingeniería en Energía, Universidad Politécnica de Chiapas, Suchiapa, México

3 CONACYT – Instituto de Investigación e Innovación en Energías Renovables, Universidad de Ciencias y Artes de Chiapas, Tuxtla Gutiérrez, Chiapas, México

*Address all correspondence to: heber.vilchis@unicach.mx

References

[1] Nakamura S, Fasol G. The Blue Laser Diode. 1st ed. Berlin Heidelberg, Berlin: Springer-Verlag; 1997

[2] Hernández-Gutiérrez CA, Morales-Acevedo A, Cardona D, Contreras-Puente G, López-López M. Analysis of the performance of $In_xGa_{1-x}N$ based solar cells. SN Applied Science. 2019; **1**(628):1-7. DOI: 10.1007/S42452-019-0650-X/

[3] Watson S, Gwyn S, Viola S, Giuliano G, Slight TJ, Stanczyk S, et al. InGaN/GaN laser diodes and their applications. In: 20th Int. Conf. Transparent Opt. Networks. Bucharest, Romania: IEEE Computer Society; 2018. pp. 1-4. DOI: 10.1109/ICTON.2018.8473864

[4] Wu J. When group-III nitrides go infrared: New properties and perspectives. Journal of Applied Physics. 2009;**106**:011101. DOI: 10.1063/1.3155798

[5] Papadomanolaki E, Bazioti C, Kazazis SA, Androulidaki M, Dimitrakopulos GP, Iliopoulos E. Molecular beam epitaxy of thick InGaN (0001) films: Effects of substrate temperature on structural and electronic properties. Journal of Crystal Growth. 2016;**437**:20-25. DOI: 10.1016/j.jcrysgro.2015.12.012

[6] Ukolov DS, Chirkov NA, Mozhaev RK, Pechenkin AA. Radiation hardness evaluation of LEDs based on InGaN, GaN and AlInGaP heterostructures. In: 2019 IEEE 31st Int. Conf. Microelectron. MIEL 2019 – Proc., Nis, Serbia. 2019; pp. 197-200. DOI: 10.1109/MIEL.2019.8889651

[7] Gao H, Ye H, Yu Z, Zhang Y, Liu Y, Li Y. Point defects and composition in hexagonal group-III nitride monolayers: A first-principles calculation. Superlattices and Microstructures. 2017;**112**:136-142. DOI: 10.1016/j.spmi.2017.09.022

[8] Tessarek C, Figge S, Aschenbrenner T, Bley S, Rosenauer A, Seyfried M, et al. Strong phase separation of strained InxGa1-xN layers due to spinodal and binodal decomposition: Formation of stable quantum dots. Physical Review B. 2011; **83**:115316. DOI: 10.1103/PHYSREVB.83.115316/FIGURES/14/MEDIUM

[9] David A, Grundmann MJ. Influence of polarization fields carrier lifetime and recombination rates in InGaN-based light-emitting diodes. Applied Physics Letters. 2010;**97**:3. DOI: 10.1063/1.3462916

[10] Chichibu SF, Onuma T, Aoyama T, Nakajima K, Ahmet P, Chikyow T, et al. Recombination dynamics of localized excitons in cubic $In_xGa_{1-x}N$/GaN multiple quantum wells grown by radio frequency molecular beam epitaxy on 3C–SiC substrate. Journal of Vacuum Science and Technology B. 2003;**21**:1856-1862

[11] Vilchis H, Compeán-García VD, Orozco-Hinostroza IE, López-Luna E, Vidal MA, Rodríguez AG. Complex refractive index of $In_xGa_{1-x}N$ thin films grown on cubic (100) GaN/MgO. Thin Solid Films. 2017;**626**:55-59. DOI: 10.1016/J.TSF.2017.02.016

[12] Jian S-R, Jang JS-C, Lai Y-S, Yang P-F, Yang C-S, Wen H-C, et al. Mechanical properties of InGaN thin films deposited by metal-organic chemical vapor deposition. Materials Chemistry and Physics. 2008;**109**:360-364

[13] Kim DJ, Moon YT, Song KM, Lee IH, Park SJ. Effect of growth pressure on

indium incorporation during the growth of InGaN by MOCVD. Journal of Electronic Materials. 2001;**30**:99-102. DOI: 10.1007/s11664-001-0107-y

[14] Romanyuk YE, Kreier D, Cui Y, Man K, Ager JW, Leone SR. Molecular beam epitaxy of InGaN thin films on Si (111): Effect of substrate nitridation. Thin Solid Films. 2009;**517**:6512-6515. DOI: 10.1016/j.tsf.2009.03.207

[15] Turski H, Siekacz M, Sawicka M, Cywinski G, Krysko M, Grzanka S, et al. Growth mechanism of InGaN by plasma assisted molecular beam epitaxy. Journal of Vacuum Science and Technology B. 2011;**29**:3-8. DOI: 10.1116/1.3590932

[16] Johnson RW, Hultqvist A, Bent SF. A brief review of atomic layer deposition: From fundamentals to applications. Materials Today. 2014;**17**:236-246. DOI: 10.1016/J.MATTOD.2014.04.026

[17] Nepal N, Anderson VR, Hite JK, Eddy CR. Growth and characterization of III-N ternary thin films by plasma-assisted atomic layer epitaxy at low temperatures. Thin Solid Films. 2015; **589**:47-51. DOI: 10.1016/j.tsf.2015.04.068

[18] Leskelä M, Ritala M. Atomic layer deposition chemistry: Recent developments and future challenges. Angewandte Chemie – International Edition. 2003;**42**:5548-5554. DOI: 10.1002/anie.200301652

[19] Akter N. Design, and simulation of indium gallium nitride multijunction tandem solar cells. International Journal of Research Engineering and Technology. 2014;**3**:315-321. DOI: 10.15623/ijret.2014.0301056

[20] Rühle S. Tabulated values of the Shockley-Queisser limit for single-junction solar cells. Solar Energy. 2016; **130**:139-147. DOI: 10.1016/j.solener.2016.02.015

[21] Satoh A. Outline of molecular simulation and microsimulation methods. In: Introd. to Pract. Mol. Simul. 1st ed. Burlington, MA, USA: Elsevier; 2011; pp. 1-27. DOI: B978-0-12-385148-2.00001-X

[22] Gunn Lee J. Computational Materials Science an Introduction. 2nd ed. Boca Ratón: CRC Press Taylor and Francis Group; 2017

[23] Stillinger FH, Weber TA. Computer simulation of local order in condensed phases of silicon. Physical Review B. 1985;**31**:5262. DOI: 10.1103/PhysRevB.31.5262

[24] Frenkel D, Beren S. Understanding Molecular Simulation from Algorithms to Applications. 2nd ed. San Diego: Academic Press; 2002. DOI: 10.1016/B978-0-12-267351-1.X5000-7

[25] Zhou XW, Jones RE, Chu K. Polymorphic improvement of Stillinger-Weber potential for InGaN. Journal of Applied Physics. 2017;**122**(23):235703. DOI: 10.1063/1.5001339

[26] Compeán García VD, Orozco Hinostroza IE, Escobosa Echavarría A, López Luna E, Rodríguez AG, Vidal MA. Bulk lattice parameter and band gap of cubic $In_xGa_{1-x}N$ (001) alloys on MgO (100) substrates. Journal of Crystal Growth. 2015;**418**:120-125. DOI: 10.1016/j.jcrysgro.2015.02.033

[27] Camas C, Conde JE, Vidal MA, Vilchis H. Molecular dynamics simulation of cubic InxGa(1-x)N layers growth by molecular beam epitaxy. Computational Materials Science. 2021; **193**:110387. DOI: 10.1016/j.commatsci.2021.110387

[28] Ohring M. Engineering Materials Science. 1st ed. Amsterdam: Elsevier; 1995

[29] Namsani S, Gahtori B, Auluck S, Singh Namsani JK. An interaction potential to study the thermal structure evolution of a thermoelectric material: β-Cu_2Se. Journal of Computational Chemistry. 2017;**38**:2161-2170. DOI: 10.1002/jcc.24865

[30] You Y, Hu H, Choi J-H. Atomic adsorption on monolayer Cu_2Se: A first-principles study. Physical Chemistry Chemical Physics. 2021;**23**:9814-9821. DOI: 10.1039/d1cp00169h

[31] Etesami SA, Asadi E. Molecular dynamics for near melting temperatures simulations of metals using modified embedded-atom method. Journal of Physics and Chemistry of Solids. 2018; **112**:61-72. DOI: 10.1016/j.jpcs.2017.09.001

[32] Zhou XW, Ward DK, Martin JE, van Swol FB, Cruz-Campa JL, Zubia D. Stillinger Weber potential for the II-VI elements Zn-Cd-Hg-S-Se-Te. Physical Review B Condensed Materials. 2013; **88**(8):085309. DOI: 10.1103/physrevb.88.085309

[33] Morse PM. Diatomic molecules according to the wave mechanics. II. Vibrational levels. Physics Review. 1929;**34**:57-64. DOI: 10.1103/physrev.34.57

[34] Chetrite R, Touchette H. Nonequilibrium microcanonical and canonical ensembles and their equivalence. Physical Review Letters. 2013;**111**:120601. DOI: 10.1103/physrevlett.111.120601

[35] Piela L. Ideas of Quantum Chemistry. 2nd ed. Amsterdam: Elsevier; 2006

[36] Hwang S-F, Li Y-H, Hong Z-H. Molecular dynamic simulation for Cu cluster deposition on Si substrate. Computational Materials Science. 2012; **56**:85-94. DOI: 10.1016/j.commatsci.2012.01.010

[37] Wei J, Liu HJ, Cheng L, Zhang J, Jiang PH, Liang JH, et al. Molecular dynamics simulations of the lattice thermal conductivity of thermoelectric material $CuInTe_2$. Physics Letters A. 2017;**381**:1611-1614. DOI: 10.1016/j.physleta.2017.03.011

[38] Galicia-Pimentel U, Osorio-González D, López-Lemus J. On the morse potential in liquid phase and at liquid-vapor interface. Rev. Mex. Fís. 2006;**52**(5):422 Available from: https://rmf.smf.mx/ojs/index.php/rmf/article/view/3485

[39] Rodríguez-Vázquez C, Camas C, Conde J, Vilchis H. Molecular dynamics simulation of Cu-Se interactions for CIGS solar cells growth process. In: 2021 18th International Conference on Electrical Engineering, Computing Science and Automatic Control (CCE), Mexico City. 2021. pp. 1-4. DOI: 10.1109/CCE53527.2021.9633090

Chapter 7

Modeling of Diffusion-Collection Mechanisms in Semiconductor Devices Submitted to Ionizing Radiation

Daniela Munteanu and Jean-Luc Autran

Abstract

When an ionizing particle passes through a semiconductor device, it transfers energy and generates electron-hole pairs along its path. The excess carriers are subsequently transported throughout the semiconductor's volume via ambipolar diffusion until they either recombine or are collected and extracted typically by a biased contact or a reverse-biased p-n junction. To predict the transient electrical behavior of complementary metal-oxide semiconductor (CMOS) devices and circuits when exposed to ionizing radiation and assess their soft error rate (SER), it is fundamental to accurately model these diverse physical processes. In this chapter, we present a comprehensive modeling and analysis of the diffusion and collection mechanisms of radiation-induced charges through a semiconductor device. Analytical formulations of the collected charge, collection current, and collection velocity are developed. These equations are further employed to establish an analytical formulation of the soft error rate (SER), explaining its exponential dependence on the critical charge of the circuit. This formulation also links the SER to various physical and technological parameters, as well as to the characteristics of the radiation.

Keywords: complementary metal-oxide semiconductor (CMOS), critical charge, diffusion-collection, metal-oxide-semiconductor field-effect transistor (MOSFET), radiation effects, semiconductor, single event effects (SEE), soft error rate (SER)

1. Introduction

The Earth is constantly exposed to a continuous bombardment of cosmic rays, also known as primary cosmic rays. These cosmic rays are a mixture of highly charged, energetic particles, consisting mainly of protons, helium nuclei, and heavy ions [1]. They come from various regions of deep space as well as from the Sun and arrive at Earth from all directions [2]. In the Earth's upper atmosphere, the primary cosmic rays interact with atmospheric particles. The result is the formation of extended atmospheric showers, which produces secondary particles that reach sea level. At ground

IntechOpen

level, this produces a continuous flow of energy-distributed particles. Some of these particles (such as muons, neutrons, protons, and pions) interact with the materials of electronic devices and circuits and can cause undesirable effects on their operation or performances [3]. An important category of effects consists of single-event effects (SEEs), defined as any measurable or observable change in the state, operation, or performance of a microelectronic device, element, subsystem, or system (digital or analog), resulting from the impact of a single energetic particle [4].

An SEE is always triggered by the interaction of an incident particle with the target material. During the interaction, the incident particle necessarily transfers energy to the medium by electromagnetic or nuclear processes. As a result of these processes, some or all of the energy of the incident particles is released into the medium [5].

There are three main sequential steps involved in the production of SEE in microelectronic devices. The first step is the charge deposition as a result of the particle hitting the sensitive region of the device by one of two mechanisms: direct or indirect ionization [6]. Direct ionization involves charged particles interacting primarily with the electrons of the material's atoms. A large number of excited energetic electrons (delta rays), usually energetic enough to ionize other atoms, are produced by the ionization mechanism. These electrons produce a cascade of secondary electrons, which lose their energy and create electron-hole pairs along the path of the particle. In this way, the deposited energy is essentially converted into electron-hole pairs. The energy required to create an electron-hole pair depends on the band gap of the material. In a silicon substrate, one electron-hole pair is created for every 3.6 eV of energy lost by the particle. Other particles (e.g., neutrons) do not ionize the matter they pass through and do not interact directly with the material. However, these particles can produce SEEs because they are likely to undergo nuclear reactions with atoms in the material. This mechanism is known as indirect ionization. The result of these nuclear reactions is the production of recoils and fragments that, like the charged particles, can deposit energy along their path by direct ionization [6].

The second step in the production of SEEs is the transport of the charge produced in the dense column of electron-hole pairs. Charges are transported by two main mechanisms: the drift of the charge in the regions subjected to an electric field and the ambipolar diffusion of the charge in the neutral zones. Finally, carriers generated in the electron-hole pair column can undergo an additional physical mechanism: carrier recombination with other mobile carriers in the lattice.

The third step is the collection of transported charges by elementary structures in the device (such as biased contacts or reverse-biased junctions). This charge collection process causes a parasitic transient current to be generated. This current is then injected into the node of the circuit affected by the particle and can cause disturbances in the circuit operation, such as SEEs.

Accurate modeling of these diverse physical processes is essential to predict the transient electrical behavior of complementary metal-oxide semiconductor (CMOS) devices and circuits when exposed to ionizing radiation and to assess their soft error rate (SER). The SER is defined as the probability that an ionizing particle will cause a transient error in the circuit that affects its operation, without causing permanent damage [3, 6]. In this chapter, we focus on the mechanisms of diffusion and collection of charges induced by energetic particles interacting with the materials of CMOS devices and circuits. A large amount of simulation and modeling studies have been carried out over the last 40 years in the field of radiation-induced diffusion-collection of charges in CMOS devices [7–21]. However, there are always challenges to be overcome concerning the calculation of the collection current and the evaluation of the circuit SER. This chapter presents a comprehensive modeling and analysis of radiation-induced charge diffusion and collection mechanisms through a semiconductor device. The aim of our development, based on the fundamental equations of diffusion-collection, is to reexamine some of the established results and to propose a new modeling of the collected charge, the collection current, and the collection velocity. Two different approaches are proposed, depending on the formalisms considered for the collection current: a pure diffusion current or a conduction current that includes a collection velocity. The equations obtained for the collected charge are then used to establish an analytical formulation of the soft error rate (SER), which explains its exponential dependence on the critical charge of the circuit. The chapter is structured as follows: In Section 2, we present the principle of the diffusion-collection model, and we recall the fundamental equations of the diffusion-collection mechanisms of the radiation-induced charges. These equations are the starting point of our modeling. Section 3 presents the diffusion-collection modeling based on the theory of diffusion. We start with the theoretical case of a point source and develop analytical expressions for the carrier density, diffusion current, and collected charge. The model accuracy is then verified using random-walk drift-diffusion (RWDD) numerical simulation. Next, we consider the more realistic case of charges that are generated along the path of an ionizing particle with a constant linear energy deposition, and we study the diffusion-collection of these charges. A linear distribution of point charges is used to emulate the particle track. An analytical formulation of the collected charge is proposed and validated through RWDD simulation. The collected charge model is then used to calculate the SER. In Section 4, the modeling of the collected charge and current is developed within a formalism that includes a collection velocity. After presenting the general case of a time- and position-dependent collection velocity, we focus on two particular cases: a time-independent collection velocity and a constant collection velocity. In the latter case, we show that the exponential form of the SER, which has long been established empirically in the literature, can be obtained analytically. It relates the SER to various physical and technological parameters and the characteristics of the radiation.

2. Principle of diffusion-collection modeling

As explained in the introduction, when a charged particle traverses through a semiconductor material, the energy it releases is transformed into electron-hole pairs. In the framework of the diffusion-collection model, these pairs are then organized into a series of localized point charges, such as discrete clusters of electrons and holes, concentrated within very tiny regions of the semiconductor material and distributed along the particle track (**Figure 1**). The number of pairs immediately after the energy deposition and pair formation is denoted as δ_n and δ_p for electrons and holes, respectively, with $\delta_n = \delta_p = \delta n_0$.

The physical mechanism through which these excess charges (electrons and holes) are transported is ambipolar diffusion, particularly in the neutral zones of the device where there is no electric field. The model then assumes that a three-dimensional (3D) spherical diffusion law in the semiconductor domain governs the transport of these excess carriers. In the case of electrons, the time and space variation of the excess electron density n_e (m^{-3}) is given by the following equation [8]:

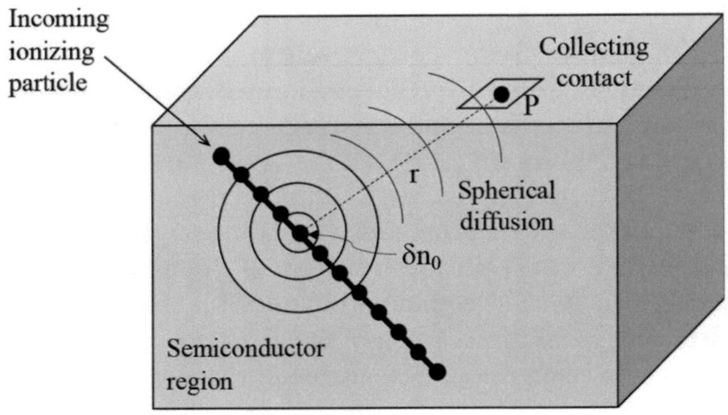

Figure 1.
Principle of the diffusion-collection modeling of radiation-induced charges.

$$\frac{\partial n_e(r,t)}{\partial t} - \frac{n_e(r,t)}{\tau} = D\nabla^2 n_e \tag{1}$$

where τ is the electron lifetime, and D is the ambipolar diffusion coefficient. D is obtained from D_n and D_p, the diffusion coefficients for electrons and holes, respectively, as follows:

$$D = \frac{2D_n D_p}{D_n + D_p} \tag{2}$$

The solution of Eq. (1) gives the excess carrier density at the time t at a distance r from the point of the initial charge deposition as follows:

$$n_e(t,r) = \frac{\delta n_0}{(4\pi Dt)^{\frac{3}{2}}} e^{\left(-\frac{r^2}{4Dt} - \frac{t}{\tau}\right)} \tag{3}$$

To maintain the analytical nature of the model, we neglect carrier recombination in Eq. (3). This is an important approximation, but it simplifies the calculations in the following to a considerable extent. The new excess carrier density equation is as follows:

$$n_e(t,r) = \frac{\delta n_0}{(4\pi Dt)^{\frac{3}{2}}} e^{-\frac{r^2}{4Dt}} \tag{4}$$

As explained above, the charge transported by ambipolar diffusion is then collected by collecting structures of CMOS devices. The charge collected at point P (**Figure 1**) by a small surface contact, which does not develop an electric field in the semiconductor, can be evaluated using Eq. (4). The charge is calculated by integrating the current density at point P over the surface of the contact and over time, from 0 to infinity. At this point, two different formalisms can be chosen to evaluate the current density: the theory of diffusion or the general electrokinetic formulation of the conduction. In the following, we present models of collected charge, collected current, and SER, using both formalisms successively.

3. Diffusion-collection modeling based on the theory of diffusion

3.1 Collected charge from a point source

We start with the theoretical case of a point source of radiation-induced charges located at $t = 0$ at the O point within the semiconductor, as illustrated in **Figure 2**. The point charge contains a number δn_0 of carrier in-excess resulting from a punctual energy transfer that creates electron-hole pairs. In the following, we detail only the case of electrons, but the transport of holes can be treated similarly. The charge $q\delta n_0$ is transported by diffusion and the variation in time and space of the excess carrier density is given by Eq. (4) if neglecting the carrier recombination process. To calculate the collected charge at the P point of the collecting contact, we assume in the following that the current density is expressed in the diffusion theory formalism.

3.1.1 Diffusion current and collected charge modeling

The diffusion current density is given by the following:

$$\overrightarrow{J_{diff}}(r,t) = qD\vec{\nabla}(n_e) \tag{5}$$

The flow of charges transported by diffusion is collected at the surface of the semiconductor by a very small area contact centered at the P point, as shown in **Figure 2**. The integration of the current density (Eq. (5)) over the contact area gives the collected current as follows:

$$I_{diff}(r,t) = \iint_{A_C} \overrightarrow{J_{diff}}(r,t)\,\overrightarrow{dS} \approx qA_C D\frac{\partial n_e(r,t)}{\partial r} \tag{6}$$

where A_C is the contact area. To avoid numerical integration in Eq. (6), we assume that the contact is sufficiently "small" compared to all other geometric dimensions. The diffusion current can also be calculated according to Eq. (6) as follows:

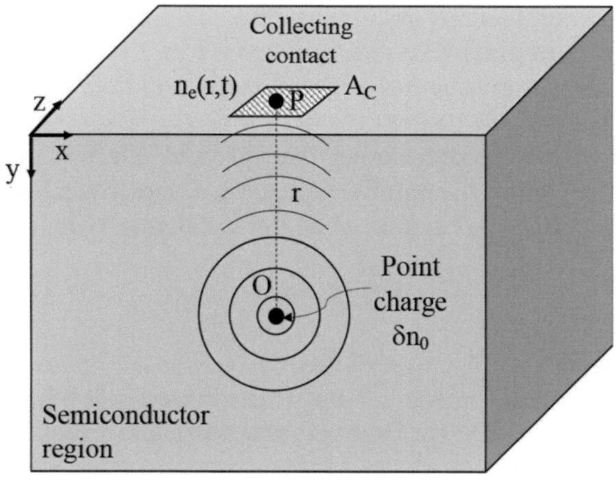

Figure 2.
Schematical illustration of the 3D spherical diffusion of a point charge and collection by a small surface collecting contact.

$$|I_{diff}(r,t)| = \frac{qA_C\delta n_0}{(4\pi D)^{\frac{3}{2}}} \times \frac{r}{2t^{\frac{5}{2}}} \times e^{-\frac{r^2}{4Dt}} = qA_C n_e(r,t) \times \frac{r}{2t} \tag{7}$$

From the collected current, the charge collected at the contact at the *P* point can be obtained by integrating Eq. (7) over time as follows:

$$q_{col}^{diff}(r,t) = \int_0^t I_{diff}(r,t')dt' = \frac{qA_C\delta n_0}{2(\pi)^{\frac{3}{2}}r^2} \times \frac{r}{2t^{\frac{5}{2}}} \times \Gamma\left(\frac{3}{2}, \frac{r^2}{4Dt}\right) \tag{8}$$

where Γ(a, x) is the upper incomplete gamma function [22]. As time tends toward infinity, the gamma function in Eq. (8) becomes equal to $\sqrt{\pi}/2$. In this case, the amount of charge collected reaches its maximum value given by the following:

$$q_{col}^{diff}(r) = q\delta n_0 \times \frac{A_C}{4\pi r^2} \tag{9}$$

The total charge collected at the *P* point, resulting from the initial charge δn_0 at the *O* point and given by Eq. (9), obviously no longer depends on time but still depends on the distance between the *O* point and the contact.

3.1.2 Verification by RWDD numerical simulation

We performed extensive numerical random walk drift-diffusion simulations to validate the diffusion current and collected charge models presented in Eqs. (7)–(9). These simulations included point sources of particles positioned at various distances from the collecting contact. We recall that the random-walk drift-diffusion (RWDD) model is a particle Monte Carlo technique. It effectively combines drift and diffusion processes using a random-walk algorithm that mimics the Brownian motion [23–25]. In the context of RWDD, the trajectory of an ionizing particle passing through the semiconductor region of devices is represented as a sequence of charge packets distributed along a linear segment. The length of this segment is equal to the range of the ionizing particle within the material. The precision of this charge discretization is guaranteed by adjusting the level of "granularity." This can be fine-tuned by choosing the packet size, which in practice typically ranges from 1 to 100 elementary charges. The transport and the recombination of the electron and hole charge packets begin immediately after the particle passes through the device. Excess carriers undergo both diffusion and drift within the preexisting "background" electric field that existed before the particle impact in the equilibrium state. The drift-diffusion motion from time t to $t + \Delta t$ of a charge packet located at $r(t)$ is calculated as follows [23–25]:

$$r(t+\Delta t) = r(t) + \mu t E_0(r(t)) + N_3(0,1)\sqrt{2D\Delta t} \tag{10}$$

where μ is the carrier ambipolar mobility, $N_3(0,1)$ is a 3D standard normal random vector (i.e., a triplet of reals between 0 and 1 distributed following the normal distribution [26]), and E_0 is the electric field before the particle strike. To account for the simplifying assumptions previously made in current and charge modeling, in the present RWDD simulation, there is no electric field, and the recombination process is neglected. A pure 3D ambipolar diffusion process is then simulated by the numerical code. During each time step of the simulation, the collected current resulting from the initial radiation-induced charge is calculated based on the transport dynamics of the minority charge packets described by Eq. (10). We use the continuity equation at the collector contact to estimate this collected current. Then, the transient current is directly computed from the number of carriers Δn that reach this contact during the time step Δt as $I(t) = q \times \Delta n/\Delta t$. More details about this simulation method and its practical implementation using an object-oriented programming language can be found in [23–25].

An illustration of the charge transport simulation after the generation of electron-hole pairs in the particle Monte Carlo RWDD approach is shown in **Figure 3**. In this simulation, we consider a radiation-induced punctual charge δn_O of 10,000 electrons located at 1 μm from a collecting contact of surface $A_C = 0.2 \times 0.2\ \mu m^2$. **Figure 3** shows a sequence of four different moments captured at t = 0, 0.1, 1, and 3 ps after the punctual charge generation, visually illustrating the progression of the RWDD charge transport process. Carriers are transported according to a 3D spherical diffusion law in all directions. Some of these carriers that either reach or cross the contact surface during the simulation run are collected at the contact.

We use RWDD simulation to verify and validate the analytical modeling of the collected charge given by Eqs. (8) and (9). **Figure 4** shows the time evolution of the

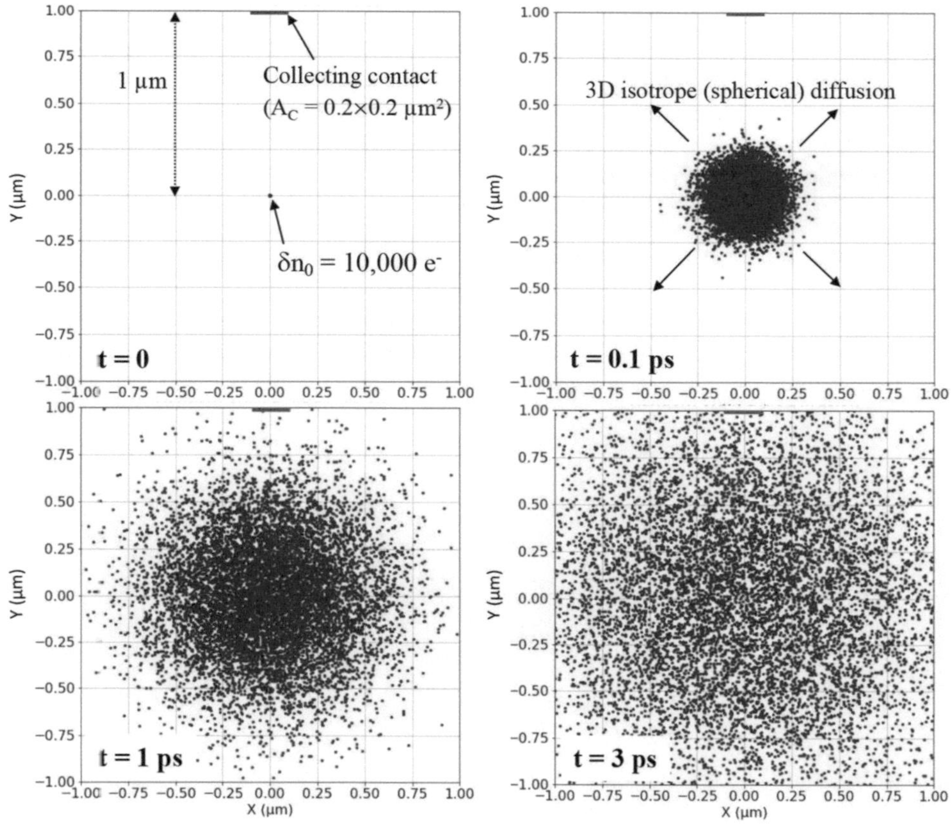

Figure 3.
Distributions of charges in a vertical section along the x-y plane (z = 0) at t = 0, 0.1, 1 and 3 ps after the deposition of 10,000 electrons at position (0, 0, 0). Charges are transported following Eq. (10) with D = 533 cm²/s, Δt = 0.1 ps, and E₀ = 0 (pure diffusion law). The surface of the collecting contact is in the x-z plane, perpendicular to the figure.*

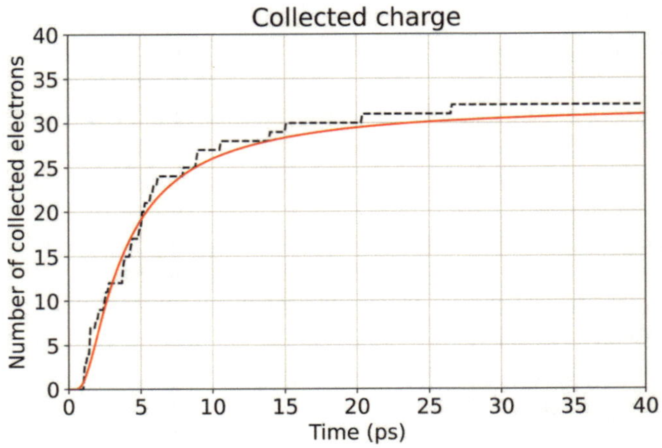

Figure 4.
*Time evolution of the charge collected by a small contact (A_C = 0.2 × 0.2 μm²) located at 1 μm from the initial deposition of 10,000 electrons (see **Figure 3**) analytically predicted by Eq. (8) and numerically computed using the Monte Carlo RWDD simulation code.*

charge collected by the contact given by Eq. (8) and by RWDD simulation with the parameters considered in **Figure 3**. We observe good agreement between the two curves shown in **Figure 4**. In addition, the maximum collected charge predicted by Eq. (9) closely matches that obtained in the RWDD simulation. This demonstrates the validity of Eqs. (8) and (9) in describing the time dependence of the collected charge. **Figure 4** also shows the stochastic nature of the RWDD curve due to the small amount of charge collected, which is about a few tens of electrons at the end of the transient.

3.2 Collected charge from a particle track

We consider now the more realistic case of charges that are generated along the path of an ionizing particle with a constant linear energy deposition. We examine a simplified scenario, as shown in **Figure 5**, in which an ionizing particle strikes the

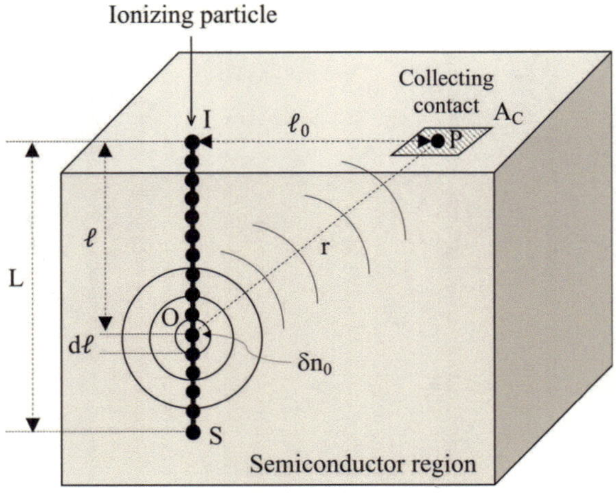

Figure 5.
Schematic representation of a particle impact in the semiconductor region of a circuit. The particle track is treated as a set of point charges δn_0 uniformly distributed along a straight segment of length L.

semiconductor surface perpendicularly. It enters the semiconductor at location I, moves through the semiconductor over a path equal to its range, and stops at location S.

The particle loses the energy ΔE energy along the [IS] segment of length L. The particle creates along its track a total number of pairs given by the following:

$$N = \frac{\Delta E}{E_{e,h}} \tag{11}$$

where $E_{e,h}$ is the energy required to create an electron-hole pair in the target semiconductor material.

The particle track is treated as a set of point charges δn_0 uniformly distributed along the straight segment [IS]. With the particle track divided in this way, the equations developed previously in the case of the diffusion from a point charge can be used here to calculate the collected charge on the contact centered at the P point. In the following, we assume that the particle has a constant LET, which means that the generated charge is deposited uniformly along its track. Under this assumption, each

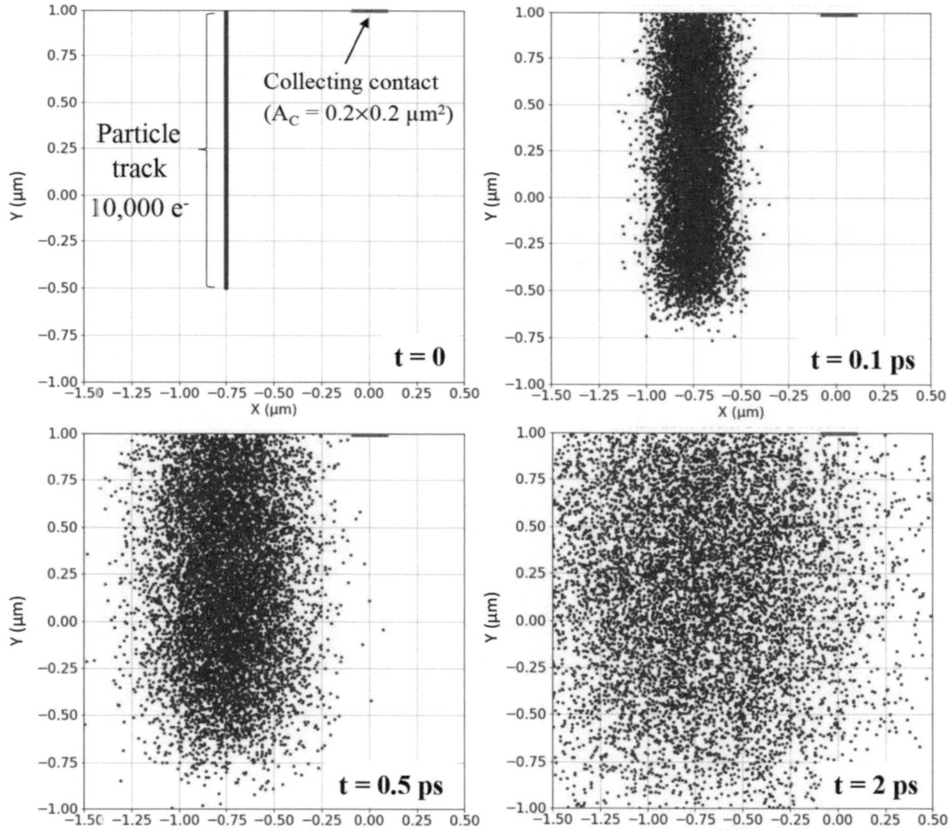

Figure 6.
Distributions of charges in a vertical section along the x-y plane (z = 0) at t = 0, 0.1, 0.5, and 2 ps after the deposition of 10,000 electrons along a 1.5-µm length track located at 0.75 µm from the contact. Other simulation parameters are the same as reported in the caption of ***Figure 3****.*

elementary segment of length $d\ell$ carries an elementary charge δn_0 considered punctual. Then, δn_0 is calculated from the total number of pairs N as follows:

$$\delta n_0 = \frac{N}{L} \times d\ell = \frac{\Delta E}{E_{e,h} L} \times d\ell \tag{12}$$

where ℓ and $d\ell$ are defined in **Figure 5**.

The total collected charge at P point, Q_{col}^{diff}, can be now calculated by integrating Eq. (9) along the entire particle track, with $r = \sqrt{\ell_0^2 + \ell^2}$ as follows:

$$Q_{col}^{diff} = \frac{q A_C N}{4\pi L} \int_0^L \frac{d\ell}{\ell_0^2 + \ell^2} \tag{13}$$

From (13), the total collected charge at P point is given by the following:

$$Q_{col}^{diff} = Q_0^{diff} \times arctan\left(\frac{L}{\ell_0}\right) \tag{14}$$

with

$$Q_0^{diff} = \frac{q A_C N}{4\pi L \ell_0} \tag{15}$$

Eq. (14) has been validated by RWDD numerical simulation. We have considered an incident particle with a track of 1.5 μm hitting the semiconductor at a distance of 0.75 μm from the contact. A sequence of four different moments captured at t = 0, 0.1, 0.5, and 2 ps after the charge generation is shown in **Figure 6**. As in the case of a point charge, this series of figures shows the transport of the radiation-induced charge by a spherical diffusion law. A part of the diffused charge is finally collected by the small

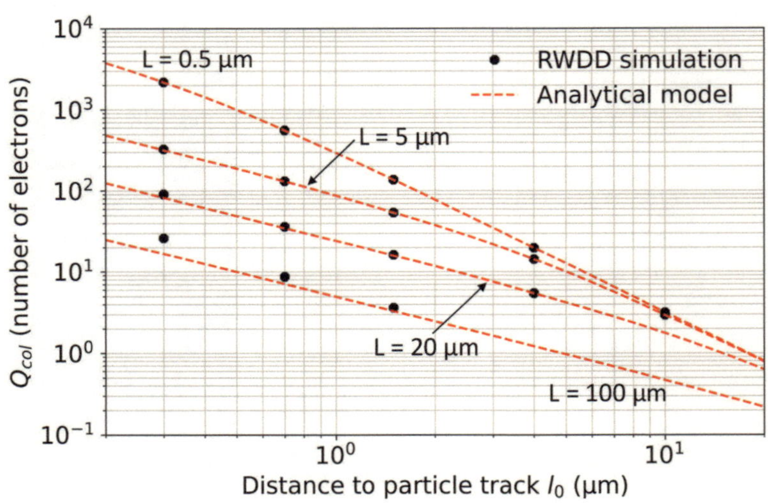

Figure 7.
*Collected charge versus the distance ℓ_0 from the collecting contact to the particle track of length L (see **Figure 5**). The collected charge is analytically calculated using Eq. (14) and numerically computed using the Monte Carlo RWDD simulation code.*

contact area. Next, we have been able to compare the numerical simulation and the analytical model of the collected charge given by Eqs. (14) with (15) through systematic RWDD simulations for different values of ℓ_0 and L. The results are shown in **Figure 7** for four values of the track length L. A very good agreement was obtained between the collected charge calculated with the analytical model and that simulated using the Monte-Carlo RWDD code, as shown in **Figure 7**.

3.3 SER modeling

To determine the SER of the circuit, we use the notion of critical charge, which is a key parameter for assessing the susceptibility to radiation of the circuit. The critical charge Q_{crit} is the minimum amount of collected charge required to trigger a transition from a logical low to a logical high state or vice versa at a sensitive node of the circuit [3, 6]. In the following, we will focus on the case where the collected charge is equal to the critical charge of the circuit. Eq. (15) shows that the collected charge is a function of the position ℓ_0 between the track and the contact. We define ℓ_{crit} as the distance ℓ_0 for which $Q_{col}^{diff} = Q_{crit}$ as follows:

$$Q_{col}^{diff} = Q_{crit} = \frac{qA_CN}{4\pi L\ell_{crit}} \times arctan\left(\frac{L}{\ell_{crit}}\right) \tag{16}$$

According to Eq. (16), a particle hitting the semiconductor at a distance ℓ_0 smaller than ℓ_{crit} will result in a collected charge greater than Q_{crit}, thus, causing a change of state of the circuit (i.e., an error). Otherwise, a particle track located at a distance greater than ℓ_{crit} will not trigger an error. Therefore, the critical surface within which the passage of an ionizing particle will trigger an error is the surface of the disk with a radius equal to ℓ_{crit} and centered on the small contact. The SER of the circuit, that is, the rate at which errors occur at the sensitive node defined by the small area contact centered at the P point, can then be calculated from the following eq. [21]:

$$SER = \pi\ell_{crit}^2F \tag{17}$$

where F is the particle flux (in $m^{-2}\ s^{-1}$).

Usually, the critical charge Q_{crit} is a known value for a given circuit, a value that can be obtained by various types of simulation such as mixed-mode, TCAD, or circuit-level simulations. Using this value in Eq. (16), the value of ℓ_{crit} and therefore of the SER can be calculated. Eq. (16) with the unknown ℓ_{crit} is a transcendental equation that can be solved numerically to obtain ℓ_{crit} but which has no analytical solution. The numerical solving of Eq. (16) can be performed using the dichotomy or Newton-Raphson methods [27].

To express the solution of Eq. (16) analytically, we use a well-known approximation for the arctan function as follows:

$$\arctan(x) \approx \frac{\pi}{2} \times \frac{x}{1+|x|} \tag{18}$$

We set

$$x = \frac{L}{\ell_{crit}} \tag{19}$$

Then, Eq. (16) becomes.

$$x^2 - Bx - B = 0 \tag{20}$$

where B is given by the following:

$$B = \frac{8L^2 Q_{crit}}{qA_C N} \tag{21}$$

The positive solution of the second-degree polynomial in Eq. (20) directly gives ℓ_{crit} as follows:

$$\ell_{crit} = \frac{2L}{B + \sqrt{B^2 + 4B}} \tag{22}$$

The SER can be then calculated from Eqs. (17) and (22) as follows:

$$SER = \pi \ell_{crit}^2 F = \frac{4\pi L^2 F}{\left(B + \sqrt{B^2 + 4B}\right)^2} \tag{23}$$

We have compared the SER given by Eq. (23) with the SER obtained as the exact solution of Eq. (16) solved numerically by the dichotomy method. **Figure 8** shows the results of this comparison for three particle track lengths. In this figure, the SER is plotted as a function of critical charge for both calculation methods. A good agreement between the analytical model and the numerical solution is obtained, in particular for large values of the critical charge.

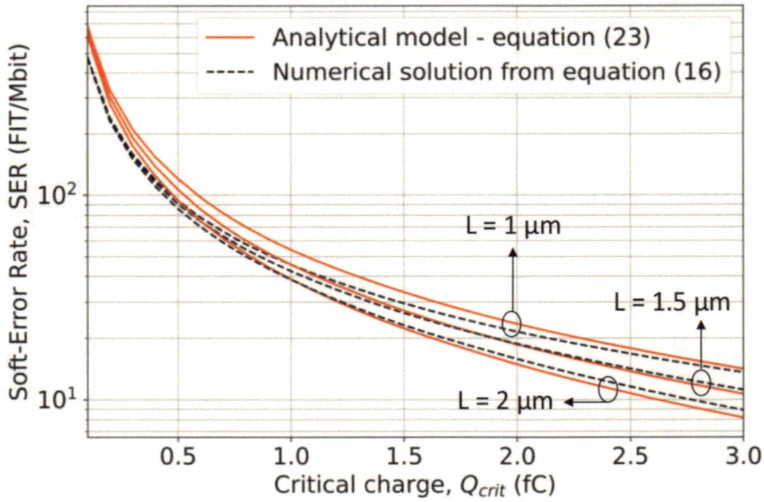

Figure 8.
Soft-error rate as a function of the critical charge calculated from the analytical model of Eq. (23) and from the numerical solving of Eq. (16). Simulation parameters are as follows: $A_C = 0.1\ \mu m^2$, $N = 1.38 \times 10^6$, $D = 61\ cm^2\ s^{-1}$, *and* $F = 10^{-3}\ cm^{-2}\ h^{-1}$.

4. Diffusion-collection based on the collection velocity

As explained before, the general kinetic formulation of conduction is a second formalism that can be used to calculate the density of the current collected at the contact coming from the charges generated along the ionizing particle track. In this section, we will use this formalism to express the current density and the collected charge, as well as the SER of the circuit. This formalism introduces the notion of collecting velocity, which leads to several approximations related to its dependence on time and space, as we will see below.

4.1 Conduction current

In the general kinetic formulation of conduction, the current density is given by the following:

$$\overrightarrow{J_{cond}} = qn\vec{v} \tag{24}$$

where n is the electron density, and v is the electron diffusion velocity. We use Eq. (24) for the carriers of density n_e transported by diffusion from the particle track to the P point. These carriers are collected by crossing the collection surface of the contact in the absence of an electric field and are uniquely transported by the diffusion process. Their diffusion velocity is therefore referred to as v_{col} for "collection velocity." The collected current is obtained by integrating Eq. (24) over the surface A_C of the collected contact as follows:

$$I_{cond}(r,t) = qA_C v_{col}(r,t) n_e(r,t) \tag{25}$$

The current density, whether expressed in diffusion theory or in the general formulation of conduction theory, must be the same. Therefore, Eqs. (5) and (24) must be equal as follows:

$$qD\vec{\nabla}(n_e) = qn_e\overrightarrow{v_{col}} \tag{26}$$

which results in the following equation for the collection velocity:

$$\overrightarrow{v_{col}} = D\frac{\vec{\nabla}(n_e)}{n_e} \tag{27}$$

From Eq. (27), v_{col} can be also rewritten as follows:

$$v_{col}(r,t) = \frac{r}{2t} \tag{28}$$

Eq. (28), which gives the general formula for the collection velocity, indicates that v_{col} depends on both the time and the distance between the particle track and the charge collection contact. Inserting the collection velocity expression (28) into Eq. (25) gives the collection current as follows:

$$I_{cond}(r,t) = qA_C n_e(r,t) \times \frac{r}{2t} \tag{29}$$

It is obvious that integrating the collection current given by Eq. (29) over time from 0 to infinity will give the same collected charge as that given by Eq. (9). In this case, the calculation of the total charge and the SER will also be the same as that described in the previous section. We therefore obtain the same collected charge using both formalisms for the current density. This is, of course, an expected result. In fact, the interest in using this second formalism based on the collection velocity lies in the fact that several approximations can be made to this collection velocity to simplify the calculation of the collected charge and of the SER. In the following, we will focus on these approximations of the collection velocity.

4.2 Time-independent, position-dependent collection velocity

A first approximation used in many studies [10, 13, 14, 19] is to include a time-independent collection velocity in the expression (25) for the collection current as follows:

$$I_{cond}(r,t) = qA_C v_{col}(r) n_e(r,t) \tag{30}$$

where $n_e(r,t)$ is given by Eq. (4). Integrating (30) over time gives the collected charge in the conduction current formalism as follows:

$$q_{col}^{cond}(r,t) = \int_0^t I_{col}(r,t')dt' = \frac{qA_C v_{col}(r)\delta n_0}{4(\pi)^{\frac{3}{2}}Dr} \times \Gamma\left(\frac{1}{2}, \frac{r^2}{4Dt}\right) \tag{31}$$

As time tends toward infinity, the gamma function in Eq. (31) becomes equal to $\sqrt{\pi}$, and the collected charge reaches its maximum value given by the following:

$$q_{col}^{cond}(r) = q\delta n_0 \times \frac{A_C v_{col}(r)}{4\pi Dr} \tag{32}$$

The amount of charge collected given by Eq. (32) must equal the amount of charge given by Eq. (9). This results in the following equation for the collection velocity:

$$v_{col}(r) = \frac{D}{r} \tag{33}$$

As expected, this collection velocity is independent of time and depends only on the position r. To simplify the calculations, Eq. (33) can be used to obtain an averaged velocity along the track as follows:

$$v_{col}^{av} = \frac{1}{L} \times \int_0^L \frac{D}{r} d\ell = \frac{D}{L} \times \int_0^L \frac{1}{\sqrt{\ell_0^2 + \ell^2}} d\ell = \frac{D}{L} arsinh\left(\frac{L}{\ell_0}\right) \tag{34}$$

The maximum collected charge given by Eq. (32) becomes:

$$q_{col}^{cond}(r) = \delta n_0 \times \frac{A_C v_{col}^{av}}{4\pi Dr} = q\delta n_0 \times \frac{A_C}{4\pi Lr} arsinh\left(\frac{L}{\ell_0}\right) \tag{35}$$

The total collected charge is obtained, similar to Eq. (13) and using (12), by integrating Eq. (35) along the entire particle track from 0 to L as follows:

$$Q_{col}^{cond} = \frac{qA_C N v_{col}^{av}}{4\pi DL} \int_0^L \frac{d\ell}{\sqrt{\ell_0^2 + \ell^2}} = \frac{qA_C N}{4\pi L^2} \times \left[arsinh\left(\frac{L}{\ell_0}\right)\right]^2 \tag{36}$$

The total collected charge is then given by the following:

$$Q_{col}^{cond} = Q_0 \times \left[arsinh\left(\frac{L}{\ell_0}\right)\right]^2 \tag{37}$$

where

$$Q_0 = \frac{qA_C N}{4\pi L^2} \tag{38}$$

Next, we calculate the SER according to the relation (17) in the same way as in Section 3.3. The length ℓ_{crit} is obtained by setting the condition $Q_{col}^{cond} = Q_{crit}$ in Eq. (37), which leads to the following equation for ℓ_{crit}:

$$\ell_{crit} = \frac{L}{sinh\left(\sqrt{\frac{Q_{crit}}{Q_0}}\right)} \tag{39}$$

The SER can be then calculated from Eqs. (17) and (39) as follows:

$$SER = \pi \ell_{crit}^2 F = \frac{\pi L^2 F}{\left[sinh\left(\sqrt{\frac{Q_{crit}}{Q_0}}\right)\right]^2} \tag{40}$$

The SER derived from Eq. (40) and that calculated by the numerical solving of Eq. (16) were compared in detail, as shown in **Figure 9**. Three values of particle track length were considered in this figure. A very good agreement between the analytical model and the numerical solution is observed over the whole range of critical charge values. This shows that assuming a time-independent collection velocity with a value averaged over the track length leads to an acceptable approximation of the SER compared to the exact numerical solution. The advantage of this assumption is that it simplifies the calculation of the SER and provides an analytical solution.

4.3 Constant collection velocity

A second approximation, more restrictive than the previous one, is to consider a constant collection velocity (a single numerical value that therefore does not depend on time or position) in the expression of the collected charge (32) [21]. We denote as v_{COL} this constant velocity. Eq. (32) becomes as follows:

$$q_{col}^{cond}(r) = q\delta n_0 \times \frac{A_C v_{COL}}{4\pi Dr} \tag{41}$$

The total collected charge is obtained by integrating Eq. (41) along the entire particle track from 0 to L as follows:

$$Q_{col}^{cond} = \frac{qA_C N v_{COL}}{4\pi DL} \int_0^L \frac{d\ell}{\sqrt{\ell_0^2 + \ell^2}} = Q_0 \times arsinh\left(\frac{L}{\ell_0}\right) \tag{42}$$

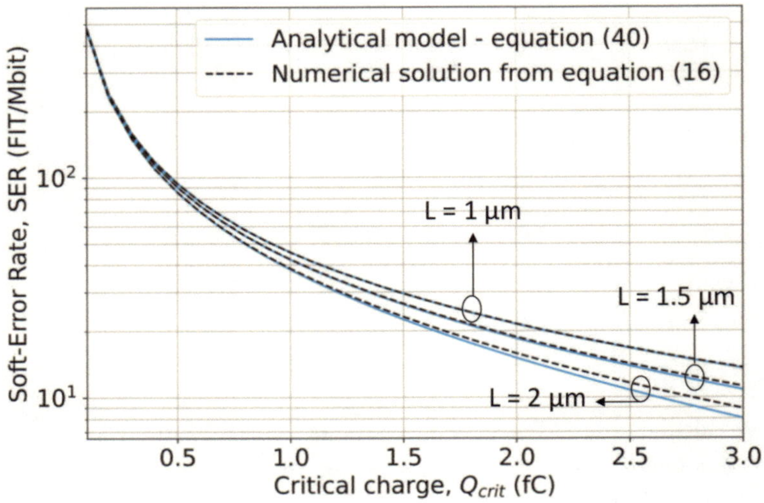

Figure 9.
Soft-error rate as a function of the critical charge calculated from the analytical model of Eq. (40) and from the numerical solving of Eq. (16). Simulation parameters are as follows: $A_C = 0.1\ \mu m^2$, $N = 1.38 \times 10^6$, $D = 61\ cm^2\ s^{-1}$, *and* $F = 10^{-3}\ cm^{-2}\ h^{-1}$.

where

$$Q_0 = \frac{qA_C N v_{COL}}{4\pi DL} \tag{43}$$

The next step is to calculate ℓ_{crit} that is obtained by setting the condition $Q_{col}^{cond} = Q_{crit}$ in Eq. (42), as explained in the previous sections. The length ℓ_{crit} is given in this case by [21] the following:

$$\ell_{crit} = \frac{L}{sinh\left(\frac{Q_{crit}}{Q_0}\right)} \tag{44}$$

Finally, the SER can be calculated from Eqs. (17) and (44) as [21]:

$$SER = \pi\ell_{crit}^2 F = \frac{\pi L^2 F}{\left[sinh\left(\frac{Q_{crit}}{Q_0}\right)\right]^2} \tag{45}$$

As explained in [21], the expression of the SER (45) including a constant value for the collection velocity makes it possible to mathematically demonstrate the exponential dependence of the SER on the critical charge, proposed more than 20 years ago by Hazucha and Svensson in [28] as follows:

$$SER = kFA_C \times e^{-\frac{Q_{crit}}{Q_S}} \tag{46}$$

where κ and Q_S are two fitting parameters. This SER formula has been established empirically based on experimental data and is used extensively in the field of SEEs in CMOS circuits. This formula was demonstrated in [21] starting from the fundamental equations of diffusion-collection. Indeed, in Eq. (44), the *sinh* function can be written as follows:

$$sinh\,(x) = \frac{e^{x} - e^{-x}}{2} \tag{47}$$

Assuming that $Q_{crit} \gg Q_0$, Eq. (44) becomes as follows:

$$\ell_{crit} = 2L \times e^{-\frac{Q_{crit}}{Q_0}} \tag{48}$$

and then the SER is given in this case by the following:

$$SER = 4\pi L^2 F \times e^{-\frac{Q_{crit}}{Q_0/2}} \tag{49}$$

Thus, we find an exponential dependence of the SER on the critical charge such as that of Eq. (44) that has been proposed empirically in the literature [28]. Eq. (49) clearly shows that this exponential dependence has a physical basis, being the result of the collection-diffusion mechanism of radiation-induced charges in the semiconductor.

In **Figure 10**, we compare the SER calculated by the exponential expression (49) with the SER obtained by the numerical solution of Eq. (16). We observe that the exponential expression matches the exact value of SER over a limited range of critical charge values (between 1.2 and 2.1 fC for L = 1.5 μm). This range varies as a function of track length, and the collection velocity v_{COL} can be considered as a fitting parameter in this case.

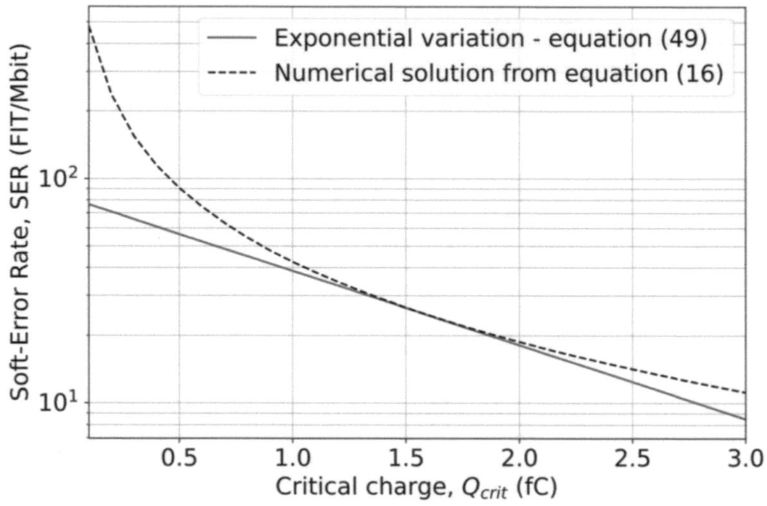

Figure 10.
Soft-error rate as a function of the critical charge calculated from the analytical model of Eq. (49) and from the numerical solving of Eq. (16). Simulation parameters are as follows: A_C *= 0.1* μm^2*,* L *= 1.5* μm*,* N *= 1.3 8 ×* 10^6*,* D *= 61* $cm^2\, s^{-1}$*,* F *=* $10^{-3}\, cm^{-2}\, h^{-1}$*, and* v_{COL} *= 4.6 ×* $10^5\, cm\, s^{-1}$ *(in Eq. (49)).*

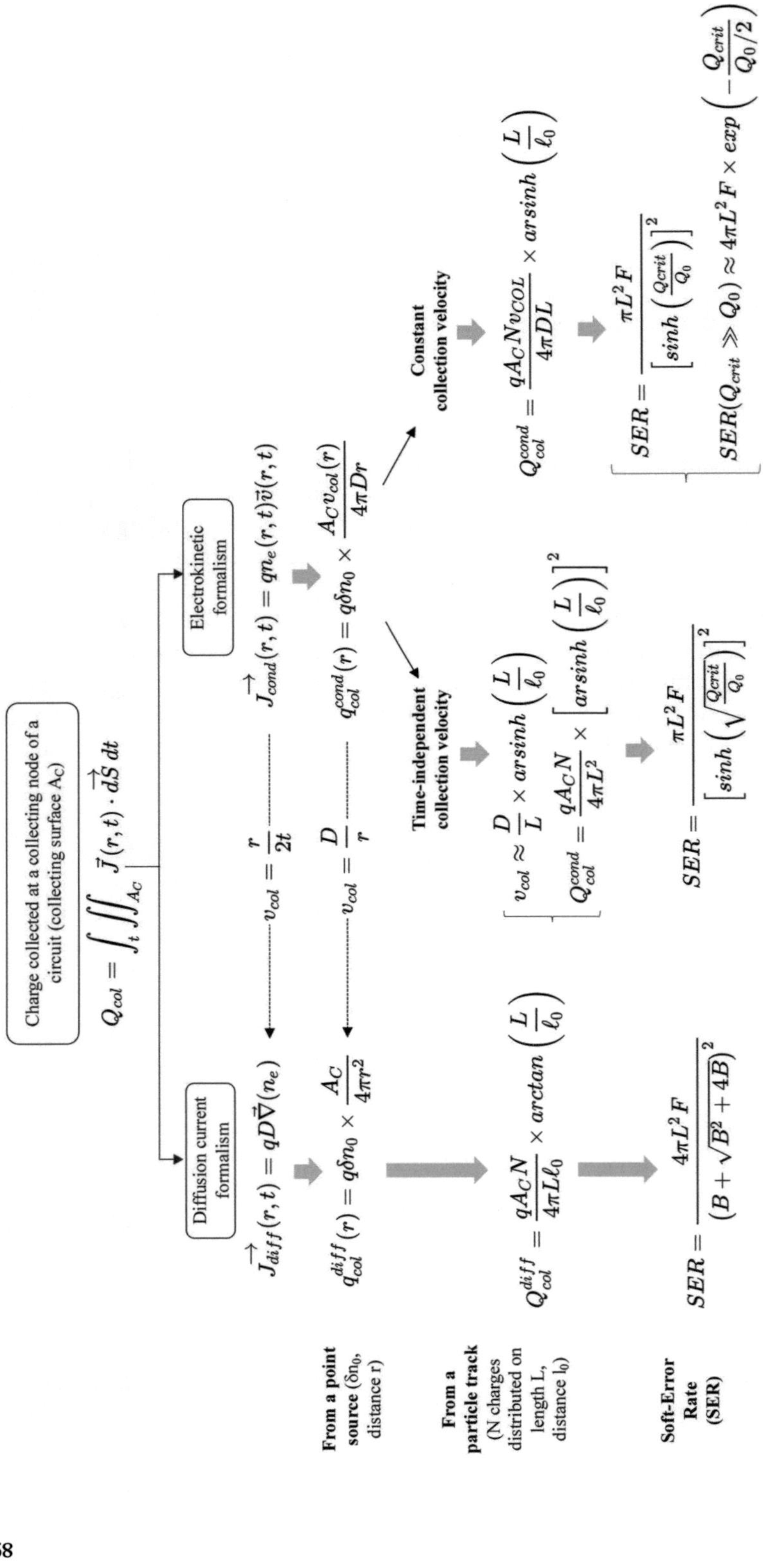

Figure 11.
Summary of the diffusion-collection model developed in this work.

5. Conclusion

In this chapter, we presented a comprehensive modeling and analysis of the mechanisms of radiation-induced charge diffusion and collection through a semiconductor device. Our development proposes a new model of the collected charge, collection current, collection velocity, and soft error rate based on the fundamental diffusion-collection equations. The successive stages of the development are summarized in **Figure 11**, including the most important equations of the model. As shown in this figure, we have considered two different formalisms for the collection current: a pure diffusion current and a conduction current that includes a collection velocity. First, we considered the theoretical case of a point charge source and developed a model for the charge collected by a small contact. Second, this charge was used to model the charge collected in the more realistic case of charges generated along the track of an ionized particle crossing the device. The equations obtained for the collected charge were then used to establish an analytical formulation of the error rate as a function of the critical charge of the circuit. Two approximations can be used to simplify the calculations and obtain analytical expressions in the case of the formalism that includes a collection velocity. A first approximation is the use of a time-independent collection velocity and the integration of an average of this velocity along the particle track into the expression for the collected charge. Our results showed that this assumption leads to an acceptable approximation of the SER compared to the exact numerical solution. A second approximation is the use of a constant collection velocity, which results in an analytical expression of the SER with an exponential dependence on the critical charge. This type of dependence has been proposed empirically in previous works. Our model showed that there is a physical basis for this exponential expression, which results from the charge collection and diffusion mechanism of the radiation-induced charges in the semiconductor.

Author details

Daniela Munteanu[1*] and Jean-Luc Autran[1,2]

1 Aix-Marseille University, CNRS, IM2NP, Marseille Cedex 20, France

2 University of Rennes, CNRS, IPR (UMR 6251), Rennes Cedex, France

*Address all correspondence to: daniela.munteanu@univ-amu.fr

References

[1] Valkovic V. Radioactivity in the Environment. Amsterdam: Elsevier; 2000

[2] Dorman LI. Cosmic Rays in the Earth's Atmosphere and Underground. Dordrecht: Kluwer Academic Publishers; 2004

[3] Autran JL, Munteanu D. Soft Errors: From Particles to Circuits. Boca Raton: CRC Press; 2015

[4] JEDEC Standard. Measurement and reporting of alpha particle and terrestrial cosmic ray-induced soft errors in semiconductor devices. N° JESD89B, Revision of JESD89A. 2021

[5] Leroy C, Rancoita PG. Principles of Radiation Interaction Matter and Detection. Singapore: World Scientific Publishing Co. Pte. Ltd; 2004

[6] Munteanu D, Autran JL. Modeling and simulation of single-event effects in digital devices and ICs. IEEE Transactions on Nuclear Science. 2008; **55**:1854-1878

[7] Kirkpatrick S. Modeling diffusion and collection of charge from ionizing radiation in silicon devices. IEEE Transactions on Electron Devices. 1979; **26**:1742-1753

[8] Messenger GC. Collection of charge on junction nodes from ion tracks. IEEE Transactions on Nuclear Science. 1982; **29**:2024-2031

[9] Edmonds L. Charged collected by diffusion from an ion track under mixed boundary conditions. IEEE Transactions on Nuclear Science. 1991;**38**:834-837

[10] Palau JM, Hubert G, Coulie K, Sagnes B, Calvet M-C, Fourtine S. Device simulation study of the SEU sensitivity of SRAMs to internal ion tracks generated by nuclear reactions. IEEE Transactions on Nuclear Science. 2001; **48**:225-231

[11] Palau JM, Wrobel R, Castellani-Coulie K, Calvet M-C, Dodd P, Sexton F. Monte Carlo exploration of neutron-induced SEU-sensitive volumes in SRAMs. IEEE Transactions on Nuclear Science. 2002;**49**:3075-3081

[12] Lambert D, Baggio J, Ferlet-Cavrois V, Flament O, Saigne F, Sagnes B, et al. Neutron-induced SEU in bulk SRAMs in terrestrial environment: Simulations and experiments. IEEE Transactions on Nuclear Science. 2004;**51**:3435-3441

[13] Merelle T, Chabane H, Palau JM, Castellani-Coulie K, Wrobel F, Saigne F, et al. Criterion for SEU occurrence in SRAM deduced from circuit and device simulations in case of neutron-induced SER. IEEE Transactions on Nuclear Science. 2005;**52**:1148-1155

[14] Wrobel F, Hubert G, Iacconi P. A semi-empirical approach for heavy ion SEU cross section calculations. IEEE Transactions on Nuclear Science. 2006; **53**:3271-3276

[15] Correas V, Saigne F, Sagnes B, Boch J, Gasiot G, Giot D, et al. Innovative simulations of heavy ion cross sections in 130 nm CMOS SRAM. IEEE Transactions on Nuclear Science. 2007;**54**:2413-2418

[16] Correas V, Saigne F, Sagnes B, Boch J, Gasiot G, Giot D, et al. Simulation tool for the prediction of heavy ion cross section of innovative 130-nm SRAMs. IEEE Transactions on Nuclear Science. 2008;**55**: 2036-2041

[17] Artola L, Hubert G, Duzellier S, Bezerra F. Collected charge analysis for a new transient model by TCAD simulation in 90 nm technology. IEEE Transactions on Nuclear Science. 2010; **57**:1869-1875

[18] Uznanski S, Gasiot G, Roche P, Tavernier C, Autran JL. Single event upset and multiple cell upset modeling in commercial bulk 65-nm CMOS SRAMs and flip-flops. IEEE Transactions on Nuclear Science. 2010;**57**:1876-1883

[19] Artola L, Hubert G, Warren KM, Gaillardin M, Schrimpf RD, Reed RA, et al. SEU prediction from SET modeling using multi-node collection in bulk transistors and SRAMs down to the 65 nm technology node. IEEE Transactions on Nuclear Science. 2011;**58**:1338-1346

[20] Martinie S, Autran JL, Sauze S, Munteanu D, Uznanski S, Roche P, et al. Underground experiment and modeling of alpha emitters induced soft-error rate in CMOS 65 nm SRAM. IEEE Transactions on Nuclear Science. 2012; **59**:1048-1053

[21] Autran JL, Munteanu D. Physics-based analytical formulation of the soft error rate in CMOS circuits. IEEE Transactions on Nuclear Science. 2023; **70**:782-791

[22] Paris RB. Incomplete gamma and related functions. In: Olver FWJ, Lozier DW, Boisvert RF, Clark CW, editors. NIST Handbook of Mathematical Functions. Cambridge: Cambridge University Press; 2010. Chapter 8

[23] Autran JL, Glorieux M, Munteanu D, Clerc S, Gasiot G, Roche P. Particle Monte Carlo modeling of single-event transient current and charge collection in integrated circuits. Microelectronics Reliability. 2014;**54**:2278-2283

[24] Glorieux M, Autran JL, Munteanu D, Clerc S, Gasiot G, Roche P. Random-walk drift-diffusion charge-collection model for reverse-biased junctions embedded in circuits. IEEE Transactions on Nuclear Science. 2014;**61**:3527-3534

[25] Autran JL, Munteanu D, Moindjie S, Saad Saoud T, Malherbe V, Gasiot G, et al. Charge collection physical Modeling for soft error rate computational simulation in digital circuits. In: Sher Akbar N, Anwar Beg O, editors. Modeling and Simulation in Engineering Sciences. Vienna: London, UKIntechOpen; 2016 Chapter 6

[26] Plante I, Cucinotta FA. Monte-Carlo simulation of particle diffusion in various geometries and application to chemistry and biology. In: Chan VWK, editor. Theory and Applications of Monte Carlo Simulations. Vienna: London, UKIntechOpen; 2013

[27] Press WH, Teukolsky SA, Vetterling WT, Flannery BP, Metcalf M. Numerical Recipes in c: The Art of Scientific Computing. 2nd ed. Cambridge: Cambridge University Press; 1992

[28] Hazucha P, Svensson C. Impact of CMOS technology scaling on the atmospheric neutron soft error rate. IEEE Transactions on Nuclear Science. 2000;**47**:2586-2594